DEDICATION

To my parents, Patrick and Janet, for their unfailing
love and for always leading by example. To all my
brothers and their families, and especially to Tony for
encouraging me to trust my instincts. And to my
three girls, love you forever.

Published by
LID Publishing Limited
The Record Hall, Studio 204,
16-16a Baldwins Gardens,
London EC1N 7RJ, UK

info@lidpublishing.com
www.lidpublishing.com

A member of:

BPR
Business Publishers Roundtable

www.businesspublishersroundtable.com

© Angus Forbes, 2019
© LID Publishing Limited, 2019

Printed in Great Britain by TJ International
ISBN: 978-1-912555-30-7

Cover design: Caroline Li
Page design: Matthew Renaudin

ANGUS FORBES

GLOBAL
PLANET
AUTHORITY

HOW WE'RE ABOUT TO SAVE THE BIOSPHERE

MADRID | MEXICO CITY | LONDON
NEW YORK | BUENOS AIRES
BOGOTA | SHANGHAI | NEW DELHI

CONTENTS

PART 2.
THE GLOBAL PLANET AUTHORITY,
WHAT IT WILL DO FOR US AND HOW WE GET ONE

PROLOGUE

Roughly speaking, between 1804 when we first numbered one billion people, and 1960 when we reached three billion, we created a lot of countries. The end of the colonial era, our population growth, combined with faster travel and communications, saw countries established the world over. It was almost like corn popping. In 1776, the USA was formed; 1810, Chile and Argentina; 1822, Brazil; 1861, Italy was unified; 1867, Canada; 1871, Bismarck unified Germany; 1901, Australia became a federation; South Africa in 1910; the Xinhai Revolution of 1911 saw the Republic of China formed; Austria and Hungary in 1918; 1932, Iraq; 1945, Indonesia; 1947, India, Pakistan, Bangladesh; 1949, the People's Republic of China; and so on and so on.

This creating of countries is called the act of national self-determination. US President Woodrow Wilson quoted in his 14 points on peace in January 1918 that 'self-determination' is not a mere phrase; it is an imperative principle of action. This was followed up with the ratification of the United Nations Charter in 1945, which placed the right of self-determination into the framework of international law and diplomacy. It states that people, based on respect for the principle of equal rights, have the right to freely choose their sovereignty with no interference.

For most of us, this era of creating countries is a bygone one unless you live in Eastern Europe in particular, where many countries were reinstated after the fall of the Berlin Wall in 1989. So, it can sound strange that a group of people can get together and just by sheer willpower create a huge organizational form like a country, but that's what was done, repeatedly.

History will show that we created most of our 200 countries when the global population was, on average, just under two billion.

Now we number just under eight billion people, we are urban, and things have changed. We now have a global problem that clearly cannot be handled by the system of independent countries and their multilateral organizations that we have created. For in the 50 years since the 1972 UN Stockholm Declaration which stated that the natural assets of Earth must be safeguarded, we have witnessed the accelerating destruction of our most valuable global asset, the biosphere. So something is structurally very wrong.

I am absolutely convinced that in order to protect the biosphere, which is essentially a global living system that knows no national boundaries, we need a specialist global authority to do the job. We need to give it powers of global regulation and global revenue collection sufficient to impose the necessary biophysical boundaries for us all.

I believe that humanity is just about to embark upon the first act of global self-determination and enter the current void in global governance to create this authority. We will do so in order to maximize our chances of survival and to uphold our greatest intergenerational responsibility, which is to ensure that we pass on a healthy bio-abundant planet. Achieving this objective is now preconditional to all other human pursuits.

We, the newly connected global populace, are now in a position to allocate part of our personal sovereignty to make this a reality. The aim of this short book is to convince you to participate in this action of global self-determination and that success is achievable.

A QUICK INTRODUCTION TO THE AUTHOR

I started talking on global governance of the biosphere in spring 2017. After one year, I realized that the idea was being well received and that people were agreeing with the concept. In addition, there were numerous requests for a book, so here we are.

The fact of the matter is that I am WWMMMM – White, Western, Male, Middle class, Middle-aged and Married (about 1% of the global population). That's really, really boring, I hear you say … yep, that's what my wife says! I wanted to get that out of the way.

I am not a scientist nor am I an academic, instead the experiences and learning I have been exposed to that have led me to passionately advocate the formation of a Global Planet Authority (GPA), come essentially in four buckets.

My forefathers fought. For many of us, members of our family from just a couple of generations ago fought in the two world wars. On my paternal side alone, my grandfather fought in World War 1, my two uncles fought in World War 2 and my father fought in the Korean War. Together with the female side of the family, they made a sacrifice to stave off what was deemed to be grave or existential risk. I think that is relevant now.

The first 20 years of my life were spent living in Adelaide, Hong Kong, Hamburg, Port Moresby, Belgrade and Auckland. My father was an Australian diplomat. I went everywhere with my parents and brothers, and I suppose I was a young global citizen. I realized early in my life the commonality of humans, that we are all just people, who appreciate politeness, respect and kindness, who try to feed and care for our families.

The second 20 years I lived in London where I worked in the stock market, at capitalism's cutting edge. I worked on big trading floors like those you might see in the movies. It was exciting, the human race and our capitalism was on the move and it reflected in the stock market as globalization took hold. I ended up specializing in consumer shares, such as those of drinks and food companies and luxury goods companies. I know how much L'Oréal spends on advertising and how a supermarket shelf is stacked so that moderately expensive products are at your eye line, for example. I understand the power of capitalist consumerism.

While I was working in the City, my interest in the environment grew, as it has done for so many of us in recent decades, and over the last ten years I have had good exposure to environmental experts, read quite a lot, started a charity in the environmental space and was the first director of the Prince's Rainforest Project, just for a very short while as planned, in London. I believe I have a reasonable grasp of the efforts taking place globally to protect the biosphere.

So, I have observed people, capitalism and governance structures around the world for my whole life, and from what I have learnt concerning the environment, I can only conclude that we are receiving a very clear message: we are losing the battle to save the planet's ability to sustain us.

There is no doubt that upon the publication of this short book I will be correctly criticized as an amateur, a non-scientist, having made some factual errors, not using the exact terminology, over simplifying, an idealist, a revolutionary, being a little disrespectful and having poor command of the English language, a criticism that will be led by my mother, who is an English teacher.

But I am happy to be branded all of those things because I believe that the formation of our first global authority, designed specifically to protect a biosphere, is 100% right for you, me, all the species on this planet and the future of humanity. And I wanted to express this firm belief in my own words.

I have written *Global Planet Authority* to be read in about three hours, as I think that if you cannot explain an idea quickly in 20 minutes, as per the Proposition chapter, and more fully in a few hours, then it is not robust or simple enough. We have quite a lot to get through in under 180 minutes.

Ready? Let's go.

SOME DEFINITIONS: GOOD TO KNOW BUT NOT ABSOLUTELY NECESSARY

The Anthropocene

An epoch dating from the commencement of significant human impact on Earth's geology and ecosystems. Scientists argue about the exact start date: it could be at the time of Columbian exchange in the 1600s, or the first industrial revolution around 1750, or when we really went global in our impact, from 1945 onwards. I use 1750.

Biosphere

The entirety of all the organisms on the planet. I just love the word biosphere (life-sphere) and I use it to mean the living planet and all the Earth systems, not the strict definition here. I hope you will excuse me using it this way.

Biophysical boundaries

The boundaries humans must remain within to allow the Earth to continue to provide sufficient ecosystem services to have a healthy biosphere and for us to survive.

Ecosystem services

The services provided by the living planet to us all: clean air, regional and global temperature, fresh water, pollination, nutrient provision, disease management etc.

GDP

Gross domestic product: the monetary measure of value of final goods and services in one year.

The Holocene

Roughly, the last 11,000 years of Earth's history: from the end of the last ice age up to the Anthropocene.

Personal sovereignty

I always imagine personal sovereignty to be a piece of paper you are holding. You tear off small sections as you create governance structures relevant to where you live, thereby saying: I cede a part of my sovereignty in order to live here and abide by these rules.

Seven geopolitical zones

I contest that the world is now roughly divided into seven geopolitical zones: North Asia, South Asia, West Asia, Africa, Europe, North America and South America.

Stratosphere

The bit of the atmosphere above the troposphere, 10-70 km up. Contains the ozone layer.

Troposphere

Lowest region of the atmosphere. From Earth's surface to about 6-10 km, contains 75% of the atmosphere's mass.

Large number reminder

They go up in three zeros. One thousand × one thousand is one million (m). One thousand × one million is one billion (bn). One thousand × one billion is one trillion (tn).

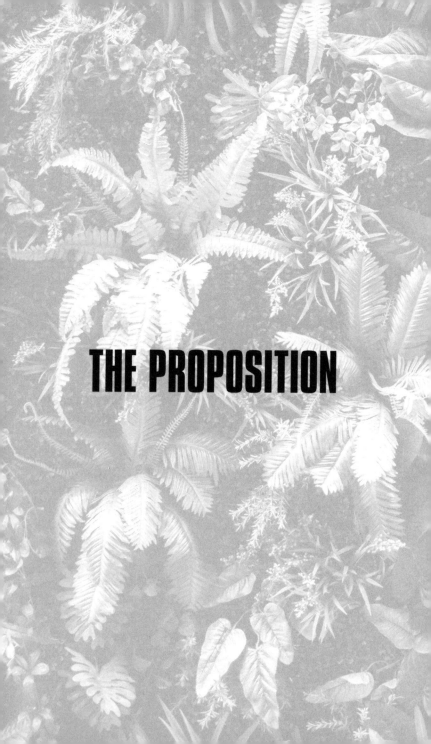

THE PROPOSITION

Two extraordinary things have just happened to the human race.

The first is the understanding that we now run, along with Mother Nature, the life-support system of our planet. The second is that we have now formed into a connected global citizenship.

For the first time in human history, we have effectively joined Mother Nature in the driving seat of the biosphere. If we wanted, we could increase carbon dioxide (CO_2) molecules in the troposphere to 600 parts per million (ppm) and heat the planet by 3°, 5° or 7°. We could use bromine for our refrigeration, which is 45 times more destructive of O_3 (ozone) molecules than chlorofluorocarbons (CFCs) and zap ourselves with UV rays. We could remove all rainforest, alter rainfall patterns, remove all species of fish and whale, dam every river, suck dry every major aquifer, scrape away all topsoil and make extinct every animal species and most plant species . If we put our minds to it, we could probably fill the Mediterranean Sea with concrete or reduce the Himalayas to rubble.

From this point on, the future of both Earth and us humans is inextricably linked due to our size and power.

About 60 years after Rachel Carson wrote *Silent Spring*, and all the subsequent environmental analysis, we now know scientifically that the biosphere is a global interconnected living system. And for the first time so are we: the combination of our population size, technology, science and industry has made us so strong that we can influence every part of the biosphere.

So, we now have part responsibility for the planet's ability to sustain life as we know it. We, yes, us humans, have to decide what the biophysical integrity of this planet will be in 2120, 2520, 3020, 4020 and thereafter. Quite a daunting realization, but one we have to accept, internalize and act upon, extremely rapidly. This new era starts today.

As it stands, we are destroying our only home. We are being told by our ecological leaders in the clearest possible terms: change our aggregate interactions with the biosphere. Now.

As Sir David Attenborough, the British ecologist, bluntly puts it:

"At present, humans are a plague on the Earth."

As Yadvinder Malhi, Professor of Ecosystem Science at the University of Oxford, explains:

"Perhaps the fundamental environmental challenge of our time, the defining characteristic of the Anthropocene, is that global industrial metabolism is of a similar size to and is surpassing the natural planetary metabolism. This leads to resource depletion, overharvesting, natural habitat loss, accumulation of metabolic waste products and atmospheric change."

Or as Stephen Emmott, the head of Computational Science at Microsoft Research in Cambridge, says even more bluntly:

"The loss of ecosystem services poses a very real threat to our survival. We can rightly call the situation we're in an un-precedented emergency. We urgently need to do – and I mean

actually do – something radical to avert a global catastrophe.
But I don't think we will. I think we are fucked."

Please excuse the language.

If you had the responsibility to protect our most valuable
asset, the biosphere, for the ultra-long term, what would
you do? I bet you wouldn't give the responsibility to 197
part-time amateurs (the nation states) who rarely talk to
each other about the subject, have continually changing
individual agendas and who have no central authority with
jurisdiction over all.

That is completely irrational, and chaos would ensue – right?

After some consideration I am confident that you would:

1. Put a specialist team in charge, whose sole job was
   to look after the asset, 24 hours a day, year after year,
   decade after decade, century after century. It would
   have our best people, a specific mandate and excellent
   internal and external governance systems.
2. Allow this team to set the necessary global biophysical
   boundaries to ensure the asset was protected and restored,
   no matter where the boundaries fell.
3. Ensure they had sufficient global power, both regulatory
   and monetary, to enforce those boundaries.

You would do these three things because they are tried and
tested as good organizational form. Specialized leadership,
clear boundaries for all and sufficient enforcement works
for us humans, because it is a fair, strong, effective and
efficient methodology.

You would recognize that in order to protect a global asset against global forces of degradation you must have a specialist dedicated **global** protectorate with sufficient powers, and that this specialist authority needs to make decisions based **on time frames different from those used by any existing human organization, i.e. 100, 500 and 1,000 years.**

Then you'd go and get on with your life.

The blame for the current predicament lies squarely with us, that is, you and me, because we have not created the right governance tool.

The nation states we have built were never designed to protect the biosphere from global attack by eight billion of us. They compete, they have other things to do like providing education and healthcare; they are not focused solely on this one vital preconditional-to-all-others global task. They don't have our best people and singular full-time focus. So we must stop sitting back and, by our inaction, continue to transfer the responsibility on to our nation states, their multilateralism or their meeting room (the UN) when they cannot provide, and will never provide, the utility that we must have.

By separating out this vital global function, we can survive, gain biophysical security and continue to progress, albeit with a much-reduced industrial metabolism.

Second, we have for the first time formed into a connected global citizenship. Just as the developers of the printing press changed the world in the last millennium by facilitating mass communication, now at the start of this millennium, the inventors of the internet, computers and smartphones have done the same. They have given us a tool to shape our future: our connectivity.

Yes, we can get caught up in the short-term divisiveness of a particular local referendum, a religious disagreement or fractious national leader, but the tide of shared values, understanding, knowledge, acceptance and respect for each other is only going one way: it is surging forward. Our art, trade, communication, media, tourism and sport reflect this.

In 2022, 32 years after Sir Tim Berners-Lee wrote the computer program HTML and gave us the World Wide Web, almost five billion of us will be connected to each other via the internet. Five billion of us only 30 seconds apart. This gives us the ability to take matters into our own hands and, for the first time, form an authority at the global level, in order to affect the whole.

This is vitally important: our 21st century technology and our newfound connectivity mean we can now take **the first action of global self-determination.**

Let's take a short trip in time. Imagine you are living in the 17th century in Europe. You live in a house where the first floor juts out a little further than the ground floor and the second floor juts out a little further still. You throw all your rubbish and effluent out of an upper-floor window into the street. As you are doing so, you look left and right and see that everyone else is doing the same.

Sure enough, the rubbish builds up and disease spreads. The town inhabitants face grave or existential risk: some could get very ill, or indeed some residents could die. However, the town inhabitants want to continue living where they are.

What do you do?

The members of the population, living only a minute or two away from each other, get together and say, "We're in real trouble here, we're all guilty and we all have a problem. Let's create an authority to deal with the problem. Let's grant it sufficient powers to do so."

The specialist authority is formed, sets a reasonable fee for its work and rapidly decides its strategy. First, it clears up the waste in the street, removing the existential risk of disease. It then bans the throwing out of rubbish and builds a maintenance system of rubbish collection. The authority maintains the health of the streets for the ultra-long term, making small adjustments over time to remain effective.

The citizens take pride in the utility they have established, give themselves high fives, or whatever they did back then, and they go about their business.

You now live at the beginning of the 21st century. The entire biosphere is being degraded. You are all guilty and you are all at risk. Experts report that humanity faces grave or indeed existential risk. The inhabitants want to continue living where they are. The members of the population realize that they are in trouble, so they **get together** and decide to create an authority to deal with the problem of global biophysical degradation.

The GPA is formed and it commandeers assets sufficient to do its work. It builds an international transaction fee (tax) system and during its first ten years it collects fees worth 6% of global GDP p.a. – which would be a 15-fold increase over current environmental spend at the national level.

The GPA rapidly decides its strategy. It shuts down ozone-depleting chemical plants, introduces a global carbon tax, places all rainforest under a global protectorate, ensures that 40% of land is held under global park status to allow biodiversity to thrive, starts planting 2bn hectares (ha) (the size of South America) of forest on already-identified degraded land, ensures at least 40% of the oceans are no take and forms global coastal parks, implements a 100% basin approach to all river systems which places the ecosystem health of the river first, oversees changes in global farming to avoid any further topsoil erosion, reduces humankind's fixed nitrogen usage by 75% and aggressively taxes landfill and petroplastics.

The authority maintains the health of the biosphere for the long term through small adjustments over time in order to remain effective.

With the GPA in place the citizens take pride in their achievement, definitely give some high fives, and go about their business.

Our first incursion into global governance must be precise. We will form the GPA and give it a clearly defined mandate: deliver biophysical integrity, that is all. Over time, we could stick AI in our heads, cure cancer, engage in regional wars, live to 150, merge countries, send people on deep space missions and it wouldn't matter. The GPA will strive, ceaselessly, to maintain the integrity of the asset. This will be the sole reason for its existence.

The GPA will have an external board, a biophysical board and an operating executive based in all of the seven geopolitical zones. The GPA must have powers over all human organizational forms: over countries, corporations and, of course, all of us. Let's be completely clear: that means it must have power over our countries and their oceans in order to do its job. It could close down companies and even move a medium-sized populace if necessary – out of a forest, for example.

While 'going global' might sound scary, it is in fact just another step on a well-trodden path. As I alluded to in my 17th century example, we have formed all sorts of utilities throughout the ages: sewage works, town councils, municipal and provincial governments, and nation states. And this is just another one. Forming the GPA does not involve the application of universal democracy to each nation state or the ceding of all nation-state sovereignty into one global government where we don't have countries anymore. It is the formation of one authority to do one job. It just happens to be our first fully supranational one.

Three key factors are already global: the asset, the destructive force and most recently, our connected population. It is only the governance structure – that is, the solution – that lags behind, because we haven't created it yet.

The GPA can now be brought into existence by holding the first ever global online voluntary vote, very likely an 'opt in' vote, whereby a quorum of us allocate a small part of our personal sovereignty in order to create the new authority that has power over the whole global population. We will vote securely using our phones.

That's it.

"What?!!" I hear you stammer incredulously. "That is terribly naive. You can't just hold a vote and create a global authority. Nation states will never agree!" To which I retort, well, we created the nation-state system and now we have to go past it and also, how have we always done it? We've voted through uprisings, through marches, through ballots, through passive opposition. But it has always effectively been a vote, a show of hands, a display of unshakable unity. An action of what is right and necessary.

I am sure that in the 1760s, if you had said to someone in London or Paris or Shanghai, the world is about to create 180 new countries, they would have said you were completely and utterly mad. Putting a global specialist in charge of the biosphere is in fact the opposite of mad, it is completely rational, urgent... and achievable. We have created all human organizational structures out of nothing, simply by human will. But we haven't built an organization capable of protecting a biosphere from our global selves.

What numbers are required to ensure this first act of global self-determination is successful? It is reasonable to argue that a global vote would need to generate a majority of just over 2.8 billion of the 5.6 billion people who are aged 18 or over (source: World Bank). But history teaches us that it is likely that fewer votes are needed, because at times of radical governance change, e.g. the forming of a nation or the removal of a racial injustice, the quorum can be formed at a lower percentage of the populace, predominantly because inactive citizens are not actually opposing the change, they just haven't got moving and are waiting for the quorum to form.

George Washington, Sun Yat-sen, Mohandas Gandhi, Nelson Mandela and Martin Luther King Jr did not wait around calmly until a neat majority was formed. The action of self-determination is more organic; it involves moral fibre, revolutionary zeal, bravery and doing what's right. A smaller quorum can take the whole with it, especially when no other effective solutions are being put forward.

With this in mind, I believe that if 1.5 billion or more of us vote in the affirmative then the GPA can be formed successfully, as 1.5 billion is bigger than either of our most populous nation states (China with a population of 1.38 billion and India with 1.33 billion), twice the size of the world's largest national election, that of India with 800 million voters, and ten times the number of votes in the next largest national elections, those of Indonesia and the USA, which have on average just 140 million voters participating.

Like all paradigm shifts it will encounter resistance; this is detailed in this book. But we will be too strong in numbers for anyone to stop. No body of peoples has ever entered the void that is global governance before and, by definition, not in such numbers. I also envisage that we, the participants in the vote, are going to let any citizen who is 13 years and older partake, as teenagers have always been active at points of change in human history. They have the most to gain from a better future.

When the GPA has been put in place and begun to lay down the necessary biophysical boundaries for our safe existence on this planet, we will in a very short space of time hardly remember a world without it. It will be a peaceful but completely radical change, and history will judge us as having effected another classic trade of utility where we all pay a small price for a big positive result, in this case a healthy, bio-abundant planet.

We will not starve; our economy will not shrivel and cease to function. Instead, our three young forces – the human population, our capitalism and consumerism – will quickly adjust, and we will thrive. By outsourcing the problem to our newly created global specialist and having the right governance structure, we can operate within clearly defined biophysical limits even with a population size of eight billion to ten billion.

It will involve hard work and sacrifice in the short term, of that there is no doubt, but I believe that we are in fact desperate to live in a world of clear biophysical boundaries, to show that we can shoulder this, our greatest intergenerational responsibility. After all, we are all just humans and all part of the biosphere. As we damage it, we damage ourselves. If we restore and look after it, we restore and look after ourselves.

Because we know the problem and we are now connected, we have in place the two preconditions for positive radical change. We are in fact 80% of the way to forming the GPA.

We have no one to ask except ourselves in order to take the last step, to vote the GPA into existence.

We must overcome our limitations by an evolutionary leap in governance.

PART 1

OUR ACCELERATING DESCENT INTO BIOPHYSICAL HELL AND HAVING THE WRONG OPERATING SYSTEM FOR A BIOSPHERE

MOTHER NATURE
AND THE HOLOCENE

I like to think of Mother Nature as sitting in a control room, perhaps the size of a large room in a house. Just as in the cockpit of a plane, there are thousands of buttons everywhere. The occasional one flashes, but Mother Nature barely glances up as a positive or negative feedback loop, already designed by her, comes into play and corrects the system back to a balanced state.

She is pretty relaxed after four billion years of evolution and, frankly, a little smug at her latest creation: the micro-era, that is the Holocene – the last 11,000 years post the last ice age.

It's been 11,000 calm years of biophysical perfection.

However, everything changed 270 years ago and in particular in the last 70 years as the human race entered the control room. The first species to ever be there – not the dinosaurs, not some aliens on the meteorite that hit 65 million years ago that caused the fifth mass extinction event in the planet's history. We are the first. And like children let loose in a sweet shop, we have grabbed at things all over the place, pushed a lot of buttons and caused a lot of damage.

After 250 years of this violent action, our scientists looked at the health of the planet's ecosystems and concluded:

"Human activities are the largest drivers of change at the planetary level.

"All of Earth's ecosystems are dramatically transformed via human actions.

"15/24 ecosystem services are degraded...

"...drivers of degradation in ecosystem services are growing in intensity."

Millennium Ecosystem Assessment, 2005

The first two sentences say: we now run the biosphere.

The third sentence says: we are doing a very bad job.

And the last sentence: we're getting even worse at it.

It is really, really important for us to grasp this and internalize the fact that we now run the biosphere, with Mother Nature.

I sometimes picture a yo-yo in my mind. I imagine that the yo-yo has been cut in half vertically.

I am holding one sphere, and sticking out of it is a little bit of wooden toggle. The sphere is the Holocene, and the toggle is the Anthropocene. Humankind now has to decide what the other sphere, yet to be affixed, is going to look like.

This is no longer about a little bit of damage done here and there. We have to change our mindset and actually manage our overall impact on Earth's systems, and plan for the very long term. Most often, success can be achieved by simply getting out of Mother Nature's way of course, due to her wonderful ability to rebalance and restore, but there will be areas where we have to actively intervene given the damage we have inflicted.

To see how much damage we have done, we need to look back at the last very short epoch, the Holocene. And to help better visualize the biosphere during this period, I have represented it using a playing field in Figure 1. It could be any four-sided playing field, a football or hockey pitch, netball or badminton court, or even a rough oval shape – it doesn't matter, of course.

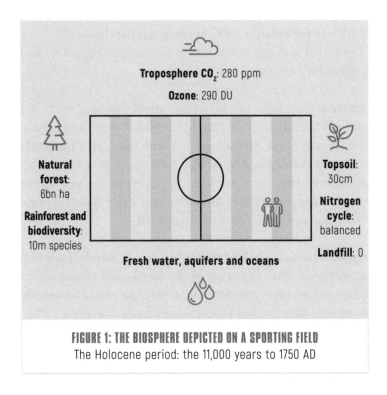

FIGURE 1: THE BIOSPHERE DEPICTED ON A SPORTING FIELD
The Holocene period: the 11,000 years to 1750 AD

Please note: Mother Nature's services are to all life, but here I'm going to describe them in just humanist terminology.

Up in the stratosphere, Mother Nature has given us ozone molecules, forming the ozone layer, which deflect UV rays and stop us being zapped by the Sun's harmful radiation. This we measure in Dobson units (DU) – Mother Nature provided 290 DU.

In the troposphere, Mother Nature has given us CO_2 molecules as one of the warming agents. The CO_2 levels in the troposphere where 280 ppm when the pyramids were built, 280 ppm during the Ming dynasty, and 280 ppm during the Renaissance. She absolutely flatlined it, as they would say. This has provided a 14°C surface temperature, ±0.5° – a wonderful average temperature for so many species, and perfect for humans.

We were given huge natural forests, about 6bn ha of them: the massive boreal forests of North America, Europe and Russia, the woodlands of Africa, together with the great rainforests in South America, Africa and Asia. Only recently it was discovered that there are literally 'rivers in the sky' above the Amazon, from the huge amounts of water that evaporate every day and then descend via rainfall.

Rainforests house half of all Earth's species. Rainforest is like a patchwork quilt made up of thousands and thousands of little squares, every one different. If you take one of these little squares of, let's say, just 10 km^2 of rainforest, there is no other like it because of the density of different species: it is unique.

Mother Nature has given life to biodiversity that we can hardly comprehend. We have identified two million species of plant, animal and insect, and we think we have discovered only about 20% of all species, so it is estimated that there could be as many as eight million species that are yet to be identified and learned about. What an extraordinary opportunity for our biomimicry and pharmacology of the future.

She has given us fresh water, rivers and aquifers as holding tanks, and stunning bountiful oceans that provide 50-70% of our oxygen through photosynthesis by phytoplankton.

It takes 500-1,000 years for just 3 cm of topsoil to be made naturally through the process of nitrogen fixation, which basically uses underground bacteria (attached to legumes) interacting with atmospheric nitrogen to produce ammonia. Mother Nature has given us 30 cm of topsoil in her most fertile valleys to grow our plants. If you hold just one handful of healthy topsoil you are holding more microorganisms – that is, more life – than the total number of humans that have ever lived.

She has her own balanced nitrogen cycle at 61m tonnes per annum (p.a.) and, of course, no landfill.

Mother Nature provided a perfect system. Complex, balanced, wild, beautiful, rich, bountiful, bio-abundant, renewing, life-giving and simply miraculous. For the last 11,000 years, she has given us Heaven on Earth.

With the Hubble telescope and all the space exploration of recent times, humankind has not found any planet that comes close to this. Well, nothing we can get to quickly, anyway.

ENTERING THE CONTROL ROOM WITH MR NEWTON

Badly paraphrased, Newton's three laws of motion are:

Law 1   An object is at rest or in steady state unless acted upon by an external force. That's the biosphere for the last 11,000 years, before we became incredibly strong.

Law 2   Upon being acted upon by an external force, the relationship *force = mass × acceleration* holds, in the same direction as the vector of the momentum. So, as we apply increasing force of degradation to a body of fixed mass, the rate of degradation accelerates. This law is holding with respect to our anthropogenic degradation of the biosphere. We keep applying more and more force.

Law 3   For every force, there is an equal and opposite force. On our current trajectory, Newton is going to be right again. But it will not be a good force, it will be something very bad for us humans, like existential risk, because Mother Nature is much stronger than us and she is being awoken from her peaceful state. We are like a really persistent mosquito in Mother Nature's room at night, that keeps coming back near her ear time and time again, and she's getting mightily annoyed... and just beginning to raise her hand in reprisal.

We started off the Anthropocene with a gentle warm-up. I call 1750 to 1950 humankind's time of **big** discovery, when we invented a lot of things on the back of our two industrial revolutions, steam and electricity; in 1909, the Haber-Bosch process miraculously gave us fixed nitrogen that could be used as a fertilizer (and ammunition, unfortunately) and with better sanitation, medicine and now more nutrients, we started living longer and procreating faster.

During this 200 year period, as our travel and tools grew better, we, well, simply put, killed anything **big**. It was like there was a global directive: if you see anything big, kill it.

In North America and Europe, we removed 80% of our boreal forest. We took the number of bison in North America from 30 million animals to just 100. Fortunately, they are back to 500,000 now thanks to some heroic conservation efforts.

The population of African elephants fell from 25 million to eight million by 1950 as we killed them to use their tusks for billiard balls and piano keys. And we made extinct the world's most populous bird, the passenger pigeon, along with the Tasmanian tiger, Steller's sea cow, the quagga, the Bali tiger and the South African Cape lion, to name just a few.

Two hundred years of big discoveries and chopping down anything big, but that was just our warm-up – a little flex of our muscles.

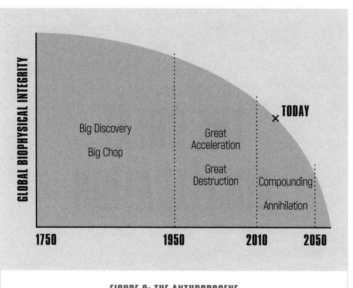

FIGURE 2: THE ANTHROPOCENE
Our non-linear destruction of the biosphere.
Being in the control room with Mr Newton

THE GREAT
ACCELERATION

POPULATION 2.5bn to 7bn	**CAPITALISM** **+** **CONSUMERISM** Go global	**FOOD** Green revolution
GDP $6tn to $70tn	**OIL** 90m barrels per day	**CARS** 75m made p.a.
PLASTIC 330m tonnes p.a.	**SUPERMARKETS** 40k stock keeping unit (SKUs)	**NET ASSETS** $350tn

FIGURE 3: THE GREAT ACCELERATION, 1950 TO 2010
An explosion in human power as we went global

From 1950, we really hit the accelerator and indeed, the period from 1950 to 2010 has been termed the Great Acceleration. It saw an explosion of human power, especially in the conversion of natural capital to human goods. Our population increased from 2.5 billion to seven billion and our nominal GDP went from $6tn p.a. to $70tn p.a. (nominal GDP refers to the reported number without an inflation adjustment, sorry to be boringly technical).

We had a green revolution in the 1950s and 1960s by adopting intensive industrial farming methods, modifying our crops to make them shorter in the stem and more yield in the head, and we added a lot of fertilizer and pesticide. During these two decades agricultural yields increased by 3.5% p.a. We fed, and continue to feed, a lot of people.

The fall of the Berlin Wall in 1989 signalled the end of large-scale communism, and capitalism won out. Capitalism and its profit motive became our global preferred operating system for the provision of human goods. And almost simultaneously came consumerism.

Today, crude oil production is 90m barrels per day, we now produce about 75m cars a year and plastic production, which was just 10m tonnes a year in 1960, reached 330m tonnes of annual production in 2010.

When you enter a large supermarket, you are typically faced with 40,000 individual line items that those in the trade call stock keeping units (SKUs). That's your batteries next to your cleaning agents, your precooked meals next to your tomatoes – 40,000 individual line items. What a provision of human goods.

And we count our assets at \$350tn, that's roughly \$200tn of urban real estate, which is why all our banks and their lending face toward housing stock, \$150tn of the rest: farms, stock-market capitalization, gold, art ... and whatever we choose to measure as value.

We have become so powerful, so quickly, it is almost unfathomable. But fathom it we must.

I believe that using a time machine, the average upper-middle-class citizen from our times could walk up to any king or queen, emperor or empress, or pharaoh of the past, and say:

"Hey there [which would startle them in the first place] ... I'm going to live to 83, have a pain-free life, not get poisoned by a snake or stabbed at the senate, and have no toothache.

I can regulate the temperature of any room I'm in, turn on a tap and drink fresh water, and buy foods from any season and most parts of the world, all year round. I can travel by an individual transport mechanism at 100km per hour, or I can go to an airport and, for just $1,000, travel anywhere in the world in just one day. Oh, and I can speak to my friends face to face on my phone, watch movies and sport and look up any piece of information in 0.5 seconds."

They'd be like: "Please, I beg you, I'll swap you Egypt, Rome or Rajasthan for that …"

But it's come at a great cost.

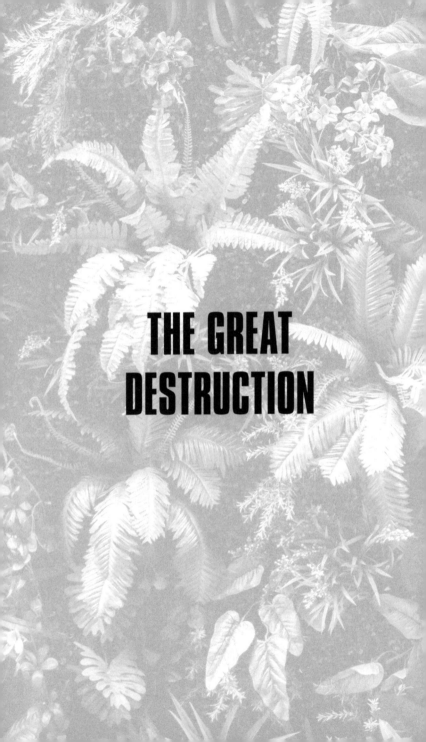

THE GREAT
DESTRUCTION

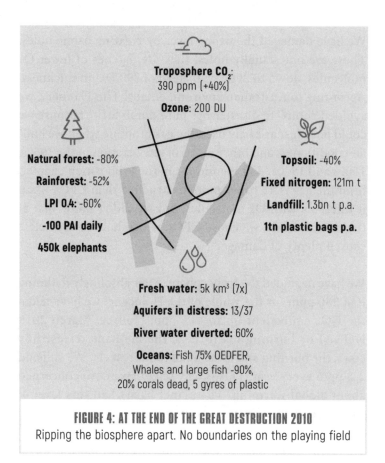

FIGURE 4: AT THE END OF THE GREAT DESTRUCTION 2010
Ripping the biosphere apart. No boundaries on the playing field

In summary, there is not one part of the biosphere that we have not materially degraded or you could say, desecrated. Desecration of our most precious asset, our life-support system.

Are we crazy?

Let's take a really quick tour, starting at the top and working anticlockwise. My apologies, but my mind's eye has always worked like that with respect to this diagram.

We have damaged the ozone layer by creating ozone holes. These are not actually holes, they are patches of fewer O_3 molecules, down to 200 DU instead of 290 DU in thickness. According to Australian environmentalist Tim Flannery, we avoided complete disaster by "pure dumb luck", because we could have just as easily used the gas bromine when we built air conditioners and other forms of refrigeration in the 1950s, 1960s and 1970s. As I mentioned in the Proposition chapter, bromine gas is 45 times more destructive than CFCs when it interacts with O_3 molecules, but we didn't know that at the time. Luckily, we chose CFCs, which have nevertheless caused plenty of damage.

We have degraded the troposphere. After absolutely flatlining it at 280 ppm for the whole of the Holocene, we have taken the CO_2 equivalent up to 412 ppm (source: March 2019 NOAA) by burning fossil fuels, the methane released by cows, the burning and clearing of rainforest etc. We still add 2-3 ppm every year. This is why scientists are so concerned about global warming. The planet hasn't had this level of CO_2 particulates for three million years.

We've cut down 80% of the world's original natural forest (source: World Resources Institute). Some 90% of continental USA's indigenous forest has been removed since 1600, according to the University of Michigan. A good portion of the world's original natural forest we have replaced with tree farming, so the net loss of forest since 1750 is one third, about 2bn ha, leaving 3.9bn ha, or 31%, of the world's land surface covered by trees.

We have been equally if not more aggressive in rainforest removal. We only started chopping down rainforest after World War 2 and we have halved that according to the UN's Food and Agriculture Organization (FAO). To get rid of half of those little unique bio-abundant squares is a disaster of unthinkable proportions. That's you and me doing that, not someone else. Got a mobile phone? There are metals in there that are mined straight from the rainforest. Or consumed some palm oil lately?

The WWF Living Planet Index (LPI in the diagram) is a density measure of animal species. The WWF monitors 3,400 animal groups and measures their numbers. The LPI was started in 1970 at a level of 1.0, and at the end of 2014 it measured 0.40. In its Global Primates Study of 2017, Duke University concluded that 75% of primates are in decline and 60% are threatened with extinction. How are the elephants faring? There were 25 million in the wild 250 years ago, in 2010 there were approximately 450,000. We are en route to zero. We kill an elephant every 26 minutes.

That's just animals. In the diagram, I give a short statistic -100PAI. It is estimated that 30 to 135 species of plants, animals and insects (PAI) are being made extinct every single day, which is almost 1,000 times nature's extinction rate.

We are now receiving dire warnings of an 'ecological Armageddon' (source: Hallmann/NAOS/IPBES) as insect numbers dramatically plunge, which scientists predict will have serious implications for all life on Earth. Biodiversity is essential for ecosystems to function.

In no period before this have there been so many plant species lost. This is important as plants need each other to survive, sharing pollinators and fighting disease together. We have no idea how the fewer plant species will fare by themselves over time.

Feeling sick yet?

We now extract seven times as much fresh water as we did in 1900. That's a really big straw. We have doubled the world's river water diverted to 60% from 30% since 1950, and we have taken the number of large dams from 2,000 to 12,000. The damming or removal of river water wreaks havoc on fragile ecosystems. We are losing amphibians at 4% p.a., as frankly they have nowhere to hop away to.

We call an aquifer that is in distress an aquifer hotspot, and 13 of the biggest 37 aquifers in the world hold that hotspot status of ill health. The most distressed is the Arabian aquifer system, followed by the Indus basin aquifer of northwest India and Pakistan, and the Murzuk-Djado basin in northern Africa. California is right up there, too. Because we have displaced so much mass, in this case water, the levels of these aquifers can now be measured by calculating changes in gravitational pull. About 18 countries, home to half of the world's population, are draining their aquifers. The world is in a fresh water crisis.

In the oceans, our actions have resulted in the removal of 90% of the numbers of whales and large fish in just 100 years – 90%! The Pacific bluefin tuna has seen population declines of 98%. Of the planet's remaining fish populations, 75% are **OEDFER**, that is **o**ver-**e**xploited, **d**epleted, **f**ully **e**xploited or **r**ecovering from exploitation. When a country states that it is fishing responsibly, like the USA or UK or Canada, within its 200 nautical mile limit, it means that it is keeping fish stocks at about one third the levels of a century ago.

Due to the warming and acidification of the oceans, 20% of our corals are dead and 20% are degraded. While we hear a lot about the Pacific gyre of plastic, we actually have five of them, and there are an estimated five trillion pieces of plastic in the oceans. If you take a teacup and scoop up any part of the ocean floor – any part – you will have four plastic fibres in your teacup. Mussels are beginning to lose their ability to cling tightly to rocks, due to the ingestion of microplastics. We have also eliminated 40% of the world's mangrove swamps.

One side of the diagram to go.

As a generalization, we can basically take it that we have got rid of 40% of topsoil. We achieve this feat by scraping at the topsoil really hard with heavy machinery; it then blows into the air and goes into the rivers and on into the oceans. Compared with Mother Nature's replenishment rate (NRR), the USA loses topsoil at ten times NRR, China and India at 30 times NRR. Humans are the **only** species on Earth to take from the soil vast quantities of nutrients and not put them back in usable form.

We have created our own, separate, nitrogen cycle. We produce 121m tonnes, compared with Mother Nature's 61m tonnes. Fixed nitrogen gets everywhere. It goes into rivers and damages the health of everything that inhabits them, it creates dead zones in oceans, it acts as a warming agent in the troposphere and even reaches the ozone layer, degrading O_3 molecules. This was the first biophysical barrier to be broken, according to scientists Johan Rockström and Will Steffen.

We put 1,300m tonnes of landfill, i.e. crap, into the ground each year. A tonne approximately equals the weight of a medium-sized car. We have 75,000 man-made chemicals, of which just 5,000 have been tested properly for environmental effects. In a recent British study, one washing load filled with polyester clothing was found to have released 700,000 microfibres of plastic into the waste water.

We have just gone upwards through the one trillion plastic bags usage mark; that's 140 plastic bags per person on Earth each year, and the number is going up. That's not pieces of plastic used. That doesn't include our plastic flip-flops or tooth-brushes, just plastic bags. Of the entire volume of plastic that is produced each year worldwide, only about 10% is recycled.

Lastly, we have a class of permanent ultrapoor people, one billion still living on less than $1 a day. Ecologists rightly argue that like all humans, desperately poor individuals are resourceful and will, for example, cut down a tree to produce charcoal in order to try and survive. Besides the moral imperative to reduce our ultrapoor to as few as possible, it is also important to do so in order to have a healthy biosphere.

CONCLUSION

Our most precious asset is under all out attack. A global web of interdependent systems that nourish each other, the miracle that is the biosphere, that allows you and me and ten million species to live, under intense global attack from a single machine made up of eight billion humans, our preferred operating system of capitalism and consumption ergo sum (with apologies to Descartes), all on a diet of carbon, fixed nitrogen, man-made chemicals and any natural resource we can get our hands on.

At the beginning of the 21st century, every one of us is now guilty of degrading the biosphere.

You can select one part, like the troposphere, and argue that rich Western nations have degraded it more in the past. But today China is the largest emitter of CO_2 and India is catching up. However, as you go around the biosphere, all nations and all peoples are guilty. For example, the country of my birth, Australia, holds the record for the highest loss of animal species per capita. In 2022, 98% of Indonesia's rainforest will have been removed (source: UN). The degree of negative human influence within European protected areas is as bad as anywhere else.

If you **still** want to argue otherwise, I'm afraid to say that you are going to live with the consequences of the destruction that has been wrought, whether you like it or not, because you live on this planet Earth.

We are all in.

If these are the effects of our new powers over just 270 years, which is a biophysical minute of time, what about the next 1,000 years or 2,000 years? What about the other half of the yo-yo? What about all our future citizens and all our fellow species? How are we going to manage our effects on the biosphere going forward?

The reason we are doing such a terrible job is because against the checklist of protecting something valuable for the ultra-long term, we are zero out of three. We have no governance of the biosphere at all.

We are structurally at fault. We have been addressing a global obligation through the prism of just the distracted nation state, which is a fatally flawed approach.

FRIEND OF FAILURE NUMBER 1: NO ONE IN CHARGE

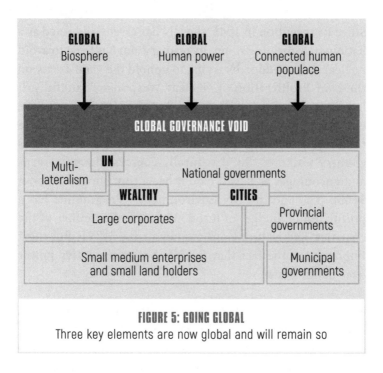

FIGURE 5: GOING GLOBAL
Three key elements are now global and will remain so

As Mother Nature decides to take her leave after four billion years of creation and 11,000 years of the perfect Holocene, she hesitates at the door to look back into her control room one final time. Humankind is busy running around pressing buttons all over the place, and she says, "Just for the record, the record of all time, which one of you clowns is in charge of doing this to my biosphere?" We look at each other blankly and then back to her. Recognizing that the answer is no one, she shakes her head in disbelief, mutters "You're even dumber than I thought" and wanders off in search of her first ever alcoholic drink.

Let's go through our organizational chart from a biophysical perspective.

Since its inception in 1945, the UN has essentially acted as a nation-state meeting room. Its primary aim has been to avoid nuclear Armageddon by trying to uphold the 1968 Treaty on the Non-Proliferation of Nuclear Weapons. Thankfully, it has done this successfully so far. But the UN does not have executive power, as evidenced by a century of its existence (adding in the League of Nations timeline) and countless country walkouts. So, it naturally coerces, shames, warns and encourages as best it can. It has a tiny budget of about $10bn p.a., and the UN Environmental Programme has a pitiful annual budget of just $335m, less than either of the charities The Nature Conservancy or the WWF. But most important is the fact that the UN has **no** executive power over nation states.

Multilateralism, more than two countries signing an agreement, is a good thing. As former British Prime Minister Macmillan said: "It is better to jaw jaw than war war." However, in the environmental space, multilateralism is like herding cats and then trying to get them to jump over the high jump.

The peace treaties of Westphalia, signed in 1648, established the legal basis of modern statehood, recognizing each country's right to rule its own territory without outside interference. At the time, Pope Innocent X referred to Westphalian settlement as "null, reprobate, and devoid of meaning for all time". Had he been referring to the effectiveness of the nation states and their multilateralism in protecting and enhancing a biosphere against our global forces, I couldn't have put it better myself.

There were plenty of early warning signs as to the lack of effectiveness of environmental multilateralism. One was the 1992 Rio Earth Summit, when 172 countries came along and **not one** binding agreement was signed. Business-class flights,

soccer on the beach, margaritas, nothing binding to sign…
I wish I had been invited to that party!

Basically, our nation states are wary of each other. Could we
go to war with them? How do I grow in power to secure my
survival? And, closely related, how do we out-compete with
them economically? A lot of it seems very old-fashioned to
me, but there we are.

Individually, governments of our nation states are busy.
They have their multitude of internal affairs to tend to,
from primary-school education to policing, economic pol-
icy, provision of infrastructure, combating terrorism and
illegal drugs, and running armies. Most do all of this on a
four- to five-year electoral cycle. Our centralist authorities,
like the Chinese, whose five-year plans are not driven by
electoral politics, also fail environmentally, which they have
readily admitted.

Corporations are designed to maximize shareholder value
and are tempted, naturally, to minimize negative externalities
that may befall them. A negative externality is a cost that is
suffered by a third party as a result of an economic transaction.
Maximizing shareholder value is their job, we designed them
that way, from the Amsterdam Stock Exchange (the world's
first modern corporation) onward.

The wealthy, a very powerful sector, are disparate and, well,
let's face it, full of individual agendas. And they know that if
they compound their returns at 7.17% p.a. for ten years, then
they have doubled their money and that's just too much fun.

Cities as governance structures are new and lightweight;
they are like national governments but without material

taxation powers. Provincial governments are isolated, and municipal governments are small and fragmented.

Small and medium-sized enterprises and small land holders are the biggest employment group in the world, but they are busy too, as President George W. Bush once sort-of-humorously and erroneously said, putting food **on** their families.

And the almost eight billion of us – well, as far as Mother Nature is concerned – seem to be capable of doing only two things: procreating and consuming.

All of these structures, organizations and peoples will play their part in restoring the biosphere, just as they have in degrading it, but there must be a dedicated specialist authority setting the rules.

As Einstein allegedly said, "We can't solve problems by using the same kind of thinking we used when we created them."

Friend of Failure Number 1 is that no one is in charge.

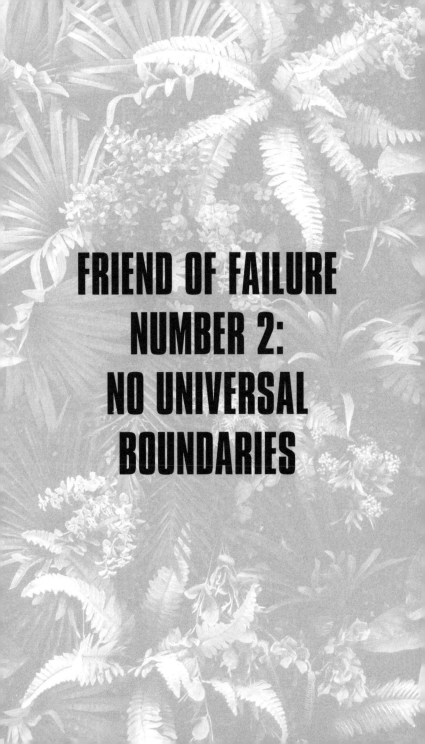

FRIEND OF FAILURE NUMBER 2: NO UNIVERSAL BOUNDARIES

Principle 2 of the UN Conference on the Human Environment held in Stockholm in 1972 stated that the natural resources of Earth must be safeguarded. The Millennium Development Goals of 2000 restated this (in goal 7a) by announcing the aim to reverse the loss of environmental resources, and the aim that by 2010 there should be a significant reduction in biodiversity loss.

Almost 50 years on from Stockholm, after millions and millions of hours of diplomacy, how are we going with that?

Our governments have signed just 300 multilateral environmental agreements (MEAs) and 250 of these 300 are as effective as sticking small Band-Aids on the backside of a raging bull. MEAs are not regulated at all, so it's all soft diplomatic stuff. You even start reading them and your eyes glaze over after 20 seconds. Here's one:

> *"The Convention on the Protection and Use of Transboundary Watercourses and International Lakes"* (just pertaining to European countries), signed in Helsinki, on 17 March 1992.

Here's another:

> *"By 2020, at the latest, biodiversity values have been integrated into national and local development and poverty reduction strategies and planning processes and are being incorporated into national accounting, as appropriate, and reporting systems."* Target 2 Aichi biodiversity targets, on 29 October 2010.

Bored yet?

Seriously, we would need about 300,000 of these to be signed in order to have any chance for them to be effective. The bull would have sticking plasters all over it, so that it would just sit down because of the collective weight of them. Indeed, the previously slow rate of signing new MEAs has dropped even further recently as nation states implicitly acknowledge that MEAs are not working due to lack of adherence.

The few that we have got right have been:

1979   Long-range Transboundary Air pollution

1986   Whaling ban (the Japanese and a few Norwegians ignore this one)

1987   Montreal Protocol banned the use of CFCs

1992   Convention on Biological Diversity: nations doubled the land in protected areas to 14%

1994   High sea drift net ban got rid of that particularly damaging form of fishing

1995   Antarctic Agreement banned drilling there

2004   Stockholm persistent organic pesticide ban

But overall it has been a disaster. Here is the E minus report card after 50 years of global boundary setting versus the targets our scientists would like to see achieved.

	TARGET	EFFECTIVE ACTION SET AT THE GLOBAL LEVEL
OZONE	290 DU	1987 Montreal Protocol
TROPOSPHERE	280 ppm	450 ppm
FOREST	2bn ha planted	350m ha by 2030
BIODIVERSITY	40%+ no pressure	1.4% no pressure
OCEANS	40-60%+ no take	2% no take
RAINFOREST	Grow	None
FRESH WATER	<4,000 km^3	None
COASTAL ECOSYSTEMS	Double	None
WETLANDS	Double	None
TOPSOIL	Preserve and increase	None
FIXED NITROGEN	-75% to 35 m tonnes	None
LANDFILL	Zero	None

OZONE LAYER: TARGET 290 DU

The 1987 Montreal Protocol protecting the ozone layer is the poster child of multilateralism. It worked, thank goodness. Our most sincere collective thanks must go to all the people who saved us. The Americans led, the Europeans, especially the British, opposed it as they tried to protect their chemical companies, but it got over the line and has been tightened over time. But this success story STILL exposes all the fragilities and pitfalls of multilateralism that will be repeated forever. Action by the nation states lagged the science, the nation states squabbled, the companies producing the offending gases fought their corner, the protocol allowed for long implementation, there was / is no central monitoring authority with power of enforcement and success was dependent on the asset being outside the nation state ownership system of land, sea or airspace. Just in case you want to be technical: developed nations will phase out CFCs, methyl bromide, carbon tetrachloride and hydrochlorofluorocarbons (HCFCs) completely by 2030. Developing countries will do so by 2040. Full recovery of the ozone layer is not expected until at least 2049 over middle latitudes and 2065 over Antarctica. We have our fingers crossed that the ozone layer will heal itself, as no one really wants to go up to where Felix Baumgartner jumped out of his Redbull canister to try and spread some O_3 molecules around.

TROPOSPHERE: TARGET 280-320 PPM OF CO_2
(Advocates Hansen, McKibben)

After 30 years of the UN trying to forge an agreement, all countries **finally** made it to the table in Paris in 2015.

How kind. Boy, did they need some coercing. To get everyone there, a $100bn annual transfer to less developed nations was agreed to, as they had correctly argued that they hadn't degraded the troposphere very much (180 of the 193 nations at the time had been responsible for less than 2% of global emissions).

Then a bizarre target was set, it was actually set in the **wrong** direction, because no big government wants to upset the carbon-based global economy. In my mind, I don't think the Intergovernmental Panel on Climate Change (IPCC), which is the scientific body involved, ever expected to be asked "Hey guys, where is the tipping point of complete disaster, because we're going to try and miss it by the smallest possible margin?" But as they were asked just that, the IPCC then held a finger to the wind, shrugged, as nobody has an exact model for it, with associated methane releases from the Siberian tundra and imploding rainforests and the subsequent release of carbon, and replied, "Well, perhaps two degrees or 450 ppm."

Where is it now, our national governments asked? At 398 ppm came the answer. Great, they said, we've got time as we are going up 2-3 ppm per year. So our national governments **agreed to maxing out at 450 ppm** and agreed that everyone could set their own national reduction targets and agreed (as always) not to police each other or fine each other. And after a big arms-in-the-air celebratory wave, they went home.

When you get dire warnings from the IPCC like the one issued in October 2018, where they effectively said, "Turn the carbon machine off now", you know why.

FOREST: TARGET REPLANT 2BN HA
(Advocates International Union for Conservation
and Nature, and University of Maryland)

Our nation states have together agreed, in the Aichi, Bonn and New York meetings, the objective of planting **350m ha by 2030**. Whether this pitiful number is achieved or not, it is forecast that we will cut down a net 1bn ha of trees, the size of the USA, for more agricultural land by 2050 – out of the current 3.9bn ha. Yes, another USA-worth of trees gone by 2050.

BIODIVERSITY: TARGET 40%+ OF LAND UNDER NO PRESSURE FROM HUMANS
(Advocates Wilson et al)

As I mentioned, protected areas have doubled to 14% of land-mass since 1992, which sounds like a really good start. But analysing these protected areas, the Wilderness Conservation Society (WCS) concluded in 2018 that only 10% of them face no pressure from humans, whereas 33% face intense pressure. So **only 1.4%** of the world's landmass is properly protected, and not surprisingly these areas are in the middle of nowhere where no one lives anyway, such as the Nullarbor Plain in Australia or within the Arctic Circle in Canada and Russia.

Observing this on the ground for many years, our environmental scientists dub the protected areas 'paper protected areas', i.e. they are only protected on paper, not in practice.

It would be remiss of me if I didn't mention the material improvement in reducing deforestation of the Brazilian Amazon rainforest achieved by its national government since 2000.

Deforestation rates are down 80%. Unfortunately, the chopping has continued apace in the countries occupying the west and north of the Amazonian basin, and satellite images indicate that Brazilian efforts have not been as effective as first thought, due to urban growth in Amazonia. And the indigenous peoples who have been given protection rights over large swathes of the rainforest are increasingly and violently threatened by nefarious characters/corporations to turn their rainforest over to them for exploitation. And there's been a change in national leader recently, and the new leader has less of a conservation bent and … and …

OCEANS: TARGET 40%-60%+ THAT IS NO TAKE
(Advocates Greenpeace, Pew, France, UK and
the University of British Columbia)

A no-take zone is an area set aside where no extractive activity is allowed. Extractive activity is any action that extracts, or removes, any resource, including fish.

Our governments have set the objective of 10% of oceans being marine protected areas (MPAs) by 2020. In 2001, the percentage was just 1%. We have moved pretty quickly on this objective and are up to 5%. But of this 5%, **only 2% is no take**, assuming (and this is a big assumption) the MPAs are policed at all. The UN's 2015 report on the state of the oceans concluded that **all 19** of humankind's negative effects on the oceans were intensifying.

And to repeat, the global objectives that our nations have set for the other biophysical assets are:

RAINFOREST	None
FRESH WATER	None
RIVERS	None
COASTAL ECOSYSTEMS	None
WETLANDS	None
TOPSOIL	None
NITROGEN	None
LANDFILL	None

At the end of the day, we are playing a global game with almost no human-imposed biophysical boundaries.

Friend of Failure Number 2.

FRIEND OF FAILURE NUMBER 3:
NUMBER 3:
NO POWER
TO ENFORCE

Cricket has the International Cricket Council, soccer has FIFA, the Olympic movement has the International Olympic Committee, netball has the International Netball Federation. Someone must be in charge to set, administer and enforce the rules of the global game; that is, there must be someone with the power of regulation. Someone must have the power of enforcement. The biosphere has nobody.

And there is no money being spent.

Our average government spends 0.4% of GDP on the environment out of the 20-35% of GDP it typically collects in taxes. To a large extent this is rational as leaders want to get re-elected and so money goes towards human-facing infrastructure and human care.

Oh no, say the Chinese, we spend 1.3% of our GDP on the environment. However, the Chinese Academy for Environmental Planning estimates that half is lost to local corruption or non-environmental projects.

Oh no, say the Europeans, we spend 0.8%. But half of that is on something titled 'waste services', and only 0.2% of GDP is spent on pollution abatement and protection of biodiversity and landscape.

In 2016, Government Spending Watch (GSW) checked the 70 nations it analyses, including India, and concluded that the spend on the environment was 0.3% of GDP and falling.

The US Environmental Protection Agency budget was forecast to be $8bn in 2018, which is 0.04% of GDP, so someone got the decimal point wrong there.

I compute the average of 0.4% of GDP to be reasonably accurate but bordering on the generous.

Spending 40 cents in $100 to protect our most valuable asset is as meaningful as trying to stop a giant moving steamroller by placing a toothpick in front of it. This is not total spend of course. The private sector is now investing and the spend on clean tech energy generation in 2018 was approximately $330 bn p.a. (another 0.4% global GDP).

Here are some recommendations made over the last few years:

- 0.5% of GDP p.a. on biodiversity alone, an eightfold increase from current levels; UN

- 2% of GDP p.a. to be spent on reducing carbon emissions (I have simplified this; his 2% figure included supply chain effects etc., but for us lay people, using a straight 2% is good enough); Lord Stern 2006

- To decarbonize the global economy, the IPCC estimates that we need to spend 6$tn p.a. (7% GDP p.a.) for ten years

- 2% of GDP p.a. on the poor and the environment, three times the current level; GSW 2017

We can also take instruction from the amounts that were spent during World War 2 in order to avoid existential risk. The UK spent 53% of its GDP p.a. for six years on the war effort, and the USA spent 41% of GDP p.a. once it entered the war in 1941. The aggressors in that confrontation, Germany and Japan, spent similar percentages. I don't know how much

Russia spent during the war, but its human toll was by far the greatest, which should be recognized by all for all time, in my humble opinion.

At the end of the day, it is pretty simple maths. To put in place the biophysical boundaries necessary, we need to spend a minimum of 5% of global GDP p.a. over a short duration of five to ten years. It could even be as high as 25 - 50% p.a. for a few years.

So we have no power of regulation and no money being spent. Friend of failure number three.

CONCLUSION

The three friends of failure are really the three friends of disaster. This situation of bad – read non-existent – governance has only led to bad outcomes and **will** only lead to bad outcomes against the global force we have unleashed. The nation-state system has been trying for 50 years and failed terribly, so do we need more evidence? Do we want to wait another 50 years so that we have 100 years of evidence of failure?

In the attempt to get to the Paris Agreement on carbon emissions, there have been more nations in and out of that process over 30 years than you can count. The Chinese crossed their arms at Copenhagen and just said 'no', so the recent bowing out by the USA is only the latest instalment in the story, and that is even with the wrong target being set.

By now, aggressive carbon taxes should have been set universally and all nations should have agreed to high petroplastic taxes and should be accurately monitoring their plastic effluent into the oceans. By now all nations should have agreed to a global target on topsoil and be paying fines if they don't meet the targets set by a central body. By now, all nations should have set a global rainforest target and be paying for it.

But they haven't even got close. Their independence, wariness and distractions will **always** stop them from doing so.

When we look back at the many futile wars in human history, historians normally conclude it was huge powerful forces, i.e. the build-up of armaments, coinciding with bad governance that led to untold human suffering in a futile conflict that didn't need to happen.

We now have all the same factors in place in our assault on the biosphere. We have huge powerful forces combined with appalling governance, which is leading to untold damage, grave or existential risk, the associated human suffering and species extinction in a futile conflict that does not need to happen.

We need to thank the many thousands of brilliant people who have dedicated themselves to conservation work or worked within the current governance structures to try to slow down the rate of destruction. People such as Muir, Vij, Hansen, Lovelock, Ngongo, Silva, Yue, Maathai, Attenborough, Galdikas, Leakey and Pandey, and all those in environmental charities, they have given us a chance. It could be even worse at this point. Hard to believe, but true.

But nothing is fit for purpose against the global forces we have unleashed. Nothing is fit for purpose for the new ultra-long era we have entered, having a direct input into the running of this planet's systems.

Some commentators believe that the economy will magically autocorrect as soon as the knowledge of environmental degradation reaches a certain point, and that will be sufficient to gain a sound biophysical future. I have been witness to the speed and number of global capital allocation decisions and the pure profit motive. I am aware of the effectiveness of advertising, the desire of the consumer, watched the efficiency with which a large corporation will move its assets around the world, and I can tell you there is not a chance of our economy doing so.

Every day, we make billions and billions of consumer decisions, and our capitalism makes millions of capital allocation decisions. This occurs every single day, except perhaps for a few hours off on Sunday afternoons. We are a pervasive unstopping global force, and opposing this we have no one in charge of the biosphere, there are almost no human human imposed biophysical boundaries and there is no power to enforce.

This situation is completely out of control.

What is that definition of insanity again? Expecting a different outcome if you don't change any of the variables?

Fasten your seatbelts. This ride is only going one way: straight down.

Welcome to compounding annihilation.

COMPOUNDING 2050

POPULATION 9.3bn	**GDP** $185tn p.a.	**GDP COMPOUND** **+2.4%**
MIDDLE CLASS 4.9bn (by 2030) **2.5x**	**FINANCIAL SERVICES** 56% emerging markets	**ELECTRICITY USE** 50bn MWH **+130%**
FOOD DEMAND **2x**	**BLUE ECONOMY** 10% mining (by 2030)	**CARS** **2x**
AIR MILES (RPKs) **3.5x**	**CHEMICAL PRODUCTION** **3x**	**PLASTIC** **2.9x**

FIGURE 6: FORECASTS FOR 2050
Compounding our human power

Our three powerful friends, the size of the human population, universal capitalism and consumerism, are young. They are all young, strong and virile. Now they want to grow and become more intense. That's fine, we have chosen them to exist, but they must not exist without global biophysical boundaries.

Let's run through the forecasts.

Our population will most likely be 9.3 billion in 2050 (UN). That's 2.3 billion more than 2010. The percentage of us living in urban centres will reach 70% by 2050, which from a capitalist point of view is good news, as ideas beget ideas and close proximity helps this accelerate.

The Organisation for Economic Co-operation and Development (OECD) forecast for the GDP growth rate of the world is 2.4% p.a. out to 2050. That means our global economy will be $185tn nominal, compared with $87tn in 2018. That is economic output every single year ... every single year! It has been proposed that the most powerful force in the universe is compounding. If that is the case, here's evidence of it.

At the heart of this growth is our new huge middle class, defined as having $10 a day income or more. The EU forecasts that another **three billion** of us will join the ranks of the middle class by just 2030, which is great, all of us should attain a moderate standard of living. This new middle class want everything I've already got: food security, energy security, a clean bathroom, an electric toothbrush etc., and it would be morally repugnant of me to argue otherwise when I have been enjoying the fruits of capitalist consumerism for the last 50 years.

PWC forecasts that 56% of financial services will be in current developing markets by 2050, up from 18% at present. What that means is that the world will send the capital required to ensure our new huge ranks of the middle class succeed in their ambition for a better economic life.

Forecasting how they will behave economically is quite straightforward. There are more econometric lines of fit and consumer S curves than you can shake a stick at. That's fancy economic talk for saying we know what our new middle classes are going to spend their money on, because other people have been there before them.

Consumer discretionary spending is 30% of income if your income is $1,500 a year, and it is 60% when your income is $15,000 a year. Our new middle class will start off purchasing shoes, low-cost manufactured goods and some furnishings and move up to more protein, financial services, healthcare and flights.

Globally, total electricity consumption will more than double by 2050 according to the World Energy Council and International Energy Agency forecasts. As it stands, not all of that is going to come from carbon – free sources, that is for sure. OPEC (biased toward oil naturally) and the International Energy Agency (a more neutral body) forecast that carbon-based fuels will still account for 70% of our total energy production, including transport use, in 2040, with oil being 28%.

Food demand will also double (source: UN). The dense grains of maize, rice, wheat and soya make up 66% of human calorific input. Yields of these four are currently increasing at just over 1% p.a., but this is before temperature variations, intense downpours and soil deterioration, are likely to reduce these yield increases over time. Doubling food demand in 32 years is 2.2% p.a. growth. It appears a small gap doesn't it – 2.2% growth in demand vs 1% yield improvement? But it is a huge gap when you are dealing with the size of our total food demand.

There's no need for me to sound a Malthusian alarm. We will simply create more agricultural land, hence the tree-removal forecast of the size of the USA by 2050. Farmland is already 45% of the land surface of this planet (source: FAO, University of Wisconsin, National Geographic). For a world whose food consumption is grinding ever higher and agricultural methods are basically unchanging, the remaining rainforest looks good to convert into new farmland, plus all of Mozambique, Tanzania and Papua New Guinea (despite the mountains).

By 2030, 10% of our total mining output will come from the oceans, with the trend only going one way. The first mining licences in international waters have already been allocated by the UN.

The World Energy Council forecasts that the number of cars in the world will double to two billion. Extending the current ten-year growth rate (Statistica) out to 2050, air miles travelled are forecast to be 3.5 times greater – that's a lot of new airports and trillions of dollars spent on aircraft. As at 2018, 82% of the world's population have not flown in an aircraft. There is huge pent-up demand.

Plastic production is set to go up by three times, i.e. it is forecast that in 2050, 1bn tonnes of plastic will be made every year versus the current 350m tonnes. Chemical production is also forecast to go up threefold (source: European Environment Agency) and the number of man-made chemical compounds to double to 150,000 (source: European Centre for Environment and Human Health).

Wave pools will be built in Snowdonia and Melbourne so we can surf inland. Large supermarkets will be built throughout Africa, nappies will be sold wherever they are currently not sold, and there will be an explosion in south–south trade. Cities are set to grow and grow in Amazonia. More dams will be built, like the one currently under construction in the middle of the Soleus Nature reserve (Tanzania), home to one of the largest populations of African elephants.

We have built a huge, compounding destruction machine, whose power and intensity is growing by the minute. Our economic caterpillar is taking larger and faster bites of our one biophysical leaf, every day.

Whoever is protecting and enhancing the biosphere, our most valuable asset, had better have their act together. They'd better have the mandate, the power, be totally focused, be applying universal biophysical boundaries and diligently enforcing them.

Oh, I forgot, there's no one there. There's nothing.

ANNIHILATION 2050

FIGURE 7: FORECASTS FOR 2050
Annihilation of the biosphere. Newton's
second law of motion is still holding

You know that sickening feeling you get when you know you are just about to witness an awful accident? The adrenaline rushes, your senses heighten, your reaction time drops to its minimum of 0.2 seconds, and you know that people are being hurt or killed at that very moment. You rush toward the scene to see if there's anything you can possibly do, although in your heart you think it is unlikely given the carnage in front of you. Soon there are screams, sirens are wailing and there are flashing lights coming around the corner. It is a terrible human-made disaster.

Figure 7, above, is what a biophysical road crash looks like: carnage and 12 red flashing lights. The forecasts all come from reputable sources with no vested interests. Don't even think about giving me the scientists-are-blowing-it-out-of-proportion-to-fund-their-research line of thought. The destruction is happening right now, as you are reading this.

In 2018 there was an article in *The Economist* warning of a few factories that are now producing the ozone-depleting chemicals dichloromethane and dichloroethane in their production of PVC and paint strippers. These gases destroy ozone.

We are adding 2-3 ppm of CO_2 equivalent in the troposphere each year, and it's not slowing down. Hence the National Oceanic and Atmospheric Administration's (NOAA) prediction of 473 ppm in 2050. I thought we were supposed to be heading back to 280 ppm!

The Hadley Centre in Exeter, UK, forecasts that at 2° C above pre-industrial levels, around 20-40% of the rainforest will disappear due to the extension of the dry season, which is already 30 days longer. Scientists refer to this dying off as a rainforest 'implosion'. When NASA predicts the future of the rainforest, it just looks at them from satellites, it doesn't try to analyse local laws or countermeasures. NASA's forecast is that it will all be gone in 50 years' time.

The Living Planet Index (WWF) is falling off a cliff. We are decimating animals, and the extinction rate is soaring. The International Union for Conservation and Nature (IUCN) forecasts that we are likely to make extinct 28% of all animal species in the next 30 years, a tragedy of literally biblical proportions. When a species is extinct … it IS extinct, gone forever.

We are now in the middle of the sixth mass extinction of species. UNESCO's IPBES (500 scientists from 137 countries) forecast that we will wipe out one million species in the next 30 years and E.O. Wilson et al forecasts that we will remove half of all the molecular life forms on the planet by 2100. After the fifth mass extinction event, 65 million years ago, it is estimated it took Mother Nature 30 million years to recreate the level of biodiversity.

Plastic in oceans is forecast to go up five times in weight by 2050 according to the Ellen MacArthur Foundation's work with the World Economic Forum. Weight is the best measure to use, not pieces, as the number of pieces of plastic in oceans will increase dramatically without us putting any more plastic in, as the existing pieces break up with waves, wind, sunlight etc. With our plastic production about to go up three times and our new huge middle class predominantly living in coastal cities, guess what happens? A fivefold increase in plastic, by weight, in the oceans by 2050.

The oceans have absorbed 85% of our CO_2 (thankfully, as otherwise the warming effect would already be much worse), which increases their acidity and bleaches, that is kills, corals. Hence the forecast that 70% of all corals will be dead in 2050 (source: NOAA and the World Resources Institute).

Fresh water use is accelerating fast, led by manufacturing use going up 400% by 2030. About three billion people will be living in areas of severe water stress (water stress is the lack of sufficient available water resources to meet the demands of water usage within a region for at least one month out of every year) in 2030 according to the most recent McKinsey/ Global Water Foundation report.

All topsoil will be gone in 60 years on current trends, warns the UN's FAO. How are we going to grow our crops?

Farming practices are forecast to be unchanged, using the same fertilizers, hence the forecast by the International Fertilizer Association (IFA) that our use of fixed nitrogen will continue to go up in line with GDP. Remember fixed nitrogen? It goes everywhere from oceans to ozone.

By 2050, the World Bank estimates that 3,000m tonnes of landfill will be put into the earth every single year, vs the 1,300m tonnes now.

Tremendous strides have been made in reducing the pecentage of the human population who are very poor, predominantly through economic growth and giving more people access to education and capitalism. However, there are still one billion of our fellow human beings living on less than $1 a day, and this is set to increase toward two billion desperate people in 2050 according to a recent UN Habitat report, as they are left behind in a poverty trap nearly always associated with bad governance.

CONCLUSION

We are going to witness the continued non-linear destruction of our life-support system. This is Newton's second law of motion still holding. This is our accelerating descent into biophysical hell.

We have all the hallmarks of imminent systemic disaster: huge powerful forces, inappropriate governance and societal behavioural inertia deriving from the sheer size, momentum and complexity of the system.

But it doesn't have to be like this. We can change our future today.

One vote to put in the necessary governance at the global level. One action of global self determination.

The following are just a few quotes, predominantly from scientists and strategists, referencing our predicament:

"History has shown that civilizations have risen, stuck to their core values and then collapsed because they didn't change. That's where we are today."

Will Steffen

"Just as in the late Roman Empire, deep stresses are rising and system resilience is declining. Governments are ever more forlornly trying to manage increasingly painful tradeoffs between people, planet and prosperity."

Thomas Homer-Dixon

"If we don't learn to live sustainably within the natural systems and limits, then we will go the same way as every other life form that failed to adapt to those changing systems and limits. Our survival as a species is not guaranteed."

Jonathon Porritt

"The truth is: the natural world ... is the most precious thing we have and we need to defend it. What humans do over the next 50 years will determine the fate of all life on the planet. Until humanity manages to sort itself out and get a coordinated view about the planet it's going to get worse and worse."

David Attenborough

"Earth is experiencing a huge episode of population declines and extirpations, which will have negative cascading consequences on ecosystem functioning and services vital to sustaining civilization."

National Academy of Sciences, 2017

"Here we are, the most clever species ever to have lived. So how is it we can destroy the only planet we have?"

Jane Goodall

"The survival of the biological support systems of the planet are at risk."

UN, 2015

"The world is still heading for catastrophic warming."

UN, 2016

"The Arctic as it is known today is almost certainly gone. On current trends the Arctic ocean will be largely ice free in summer 2040. The right response is fear."

***The Economist*, 2017**

"We have a transcendent moral obligation to all life. We are the minds of the living world whether we like it or not. We must now become its guardian and steward. A civilization able to envision God and the afterlife and embark on the exploration of space, for heaven's sake, can surely find a way to save the ecological integrity of this planet."

Edward O. Wilson

"In the Anthropocene we have to become active stewards of the Earth as a whole. Re-imagining the legal and governance constructs that people have designed to mediate the human-environment interface in the Athropocene might arguably be tantamount to a second Copernican revolution … In the words of Ayestaran, 'the first Copernican revolution placed out planet in its correct astrophysical context. A second Copernican revolution is underway that places humanity in its appropriate environmental nexus.'"

Louis Kotze, 2014

PART 2

THE GLOBAL PLANET AUTHORITY, WHAT IT WILL DO FOR US AND HOW WE GET ONE

TIME TO GO GLOBAL

The human mind is extraordinarily effective in its computational power. We can be wrong at times, even act pretty dumb at times but, boy, are we effective at searching for efficiency frontiers.

You walk into a room full of people you don't know, you're relaxed but you're alert. Your eyes are taking in 1.5 million pieces of data per second (incredible but true), and you introduce yourself to someone. Your speed of analysis is phenomenal: what are they wearing, what is the first question posed, am I safe, are they biased against me or friendly, are they with someone, who do we know in common – all in about two seconds.

Our brains are continually computing. All day, every day they are working out things: relationships, sources of food, sources of water, chances of survival, chances of procreation, employment. All the time, the brain is searching for answers and trying to be efficient.

And we haven't heard anything efficient about the protection of the biosphere.

We have heard that conservation must be on the ground, grass roots… that if we change five habits and recycle, we will save 1,203 pieces of plastic going into the ocean, that if we give just $3 a month we'll save a snow leopard (WWF giving programme).

Our brains are thinking: yes, those are factually correct and I'll participate, but something is very wrong. All the statistics are getting worse, we are ripping this biosphere apart … I haven't heard of a solution that will work and is powerful enough … there is not an efficiency frontier here.

As I have mentioned, we have resided in a pretty efficient nation state set-up for the last 100 years, created by us through numerous acts of national self-determination. But the answer no longer resides in that space, which is why our brains are saying, repeatedly and correctly, that they haven't heard the answer yet.

And those who analyse the problem agree:

"There is a disjuncture between current institutional capacity to provide public goods and the structural characteristics of a much larger-scale, global economy."
Ian Christie and Diane Warburton

"Every living system is declining, and the rate of decline is accelerating. Basically, civilization needs a new operating system and we need it within a few decades."
Paul Hawken, 2009

"The age of nations has passed. Now, unless we wish to perish, we must shake off our old prejudices and build the earth."
Pierre Teilhard de Chardin

"The dominant governance systems cannot provide the means of stopping and reversing our self-destructive behaviour. A new vision and understanding of how to govern ourselves is essential."
Cormac Cullinan

"Collectively, politicians are failing to act to maximize the chance of an acceptable outcome."

Tim Flannery

"Feeding 9 billion people while reducing pressures on other natural resources will require major changes in global governance".

The World Bank

"There is simply no machinery in today's world – to get this job done: to reach a sustainable agriculture system, and a stable temperature that we can live with – ideally close to the one we have enjoyed for the last few thousand years."

Jeremy Grantham

"We have a global ecology, a global economy and a global science – but we are still stuck with only national politics."

Yuval Harari

"When survival is threatened by seemingly insurmountable problems, an individual life form – or a species – will either die or become extinct, or rise above the limitations of its condition through an evolutionary leap."

Eckhart Tolle

"National governments can no longer control their destiny in the face of global problems."

Ayo Wahlberg

"Governance in the 21st century 'the Anthropocene' era with environmental issues at the forefront, requires different types of governance from the 20th century. We know that in many instances traditional international relations centring on the nation state no longer function well."

Peter Haas

If you get to a point where you **clearly** have the wrong operating system for the times you live in and for the known future, and you face clear risk, you have to replace it.

When it is wrong, it is just simply wrong. When you are at risk, you are at risk. Simple: there's nothing more to analyse.

After 170 years of British rule, Thomas Paine wrote *Common Sense*, his 1776 pamphlet, advocating for the formation of the USA and the removal of British rule. The 13 colonies knew they had the wrong operating system, so they acted. Thomas Paine wrote: "We have it in our power to begin the world over again."

After 268 years of Manchurian rule (part of the Qing dynasty), Sun Yat-sen led the Chinese people on their path to create a new nation. The phrase *minzu zhuyi* was his first principle, which he denoted to mean advocating self-determination, and the Republic of China was formed in 1911. "To understand is hard. Once one understands, action is easy. The whole world is one family," said Sun Yat-sen.

After 250 years of British rule, Gandhi and the Indian people acted as one, and they gained their independence in 1947. "You must not lose faith in humanity," said Gandhi.

After 81 years of apartheid, from the date of the Natives Land Act in 1913, Mandela and his followers got rid of the morally repugnant and wrong governance structure. "It always seems impossible until it's done," said Nelson Mandela.

After 270 years of anthropogenic destruction of the biosphere, we must replace the governance system.

The individuals quoted above stepped into the governance void of their time. For them, it was just as groundbreaking, just as big a step into the unknown, and yet nowadays, we simply shrug and say, "Of course the USA was formed in 1776, the Republic of China was created in 1911, India and the People's Republic of China in 1947, of course apartheid ended in 1994."

I am convinced that if Paine, Sun Yat-sen, Gandhi and Mandela were alive today, they would say to us: "GO. NOW."

And we must.

To be able to form the GPA successfully, we need to have a clear vision of what we are going to create, understand the costs and what it is going to do for us. We must believe in our ability to take this step, respect the opposition we are going to face and be able to counter it. And then we have to brave, focussed and get on and create it. We will be voting only once, to form the authority. Thereafter, the external and internal governance systems will be responsible for the running of the GPA.

THE GLOBAL PLANET
AUTHORITY

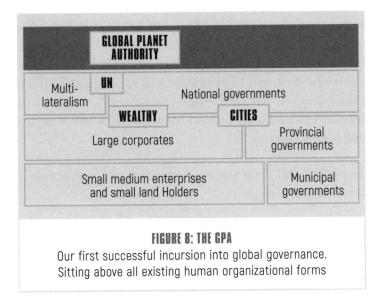

FIGURE 8: THE GPA
Our first successful incursion into global governance.
Sitting above all existing human organizational forms

The following chapters describing the GPA are my own vision. I am not a specialist in governance structures and have not sought specialist advice. The form of the GPA's structure and its fees are my own best guess; however, the recommendations of what a GPA should do with regards to biophysical boundaries come mostly from my collated research.

I wanted to paint a picture for us to visualize. I am very confident that as we build momentum and get closer to the vote, the right people will arrive to do the right job and in the right manner. Because 1.5 billion or more of us are about to act, and we are going to change the course of human history, many of our brightest minds will respond to the call and structure the best, fairest and most effective GPA.

We are very fortunate, because we have just formed into a multipolar world. This is move is timely, as it allows humanity to progress with balance and respect. I contest that we now comprise seven geopolitical zones (SGZs). The SGZs are: North Asia (China to Japan), South Asia (Myanmar to New Zealand), West Asia (The Middle East to the greater Indian subcontinent), Africa, Europe (including Russia; 80% of Russians live west of the Ural Mountains), North America and South America.

The SGZs will form the blueprint for the organizational structures of the GPA.

Sitting at the top of the GPA will be the External Board, whose responsibility will be to maintain best practice in internal governance and to report to us, the global citizens, the key metrics achieved by the GPA and their effectiveness. The External Board will have seven members, one from each of the SGZs.

The GPA will have a Biophysical Board that will be responsible for setting the goals, that is for setting the boundaries within the biosphere, without consideration of politics, economics, religion and national borders. That is because we have to yield to it, not it to us.

The Biophysical Board will be made up of 21 environmental scientists. An odd number of members always ensures a democratic outcome in decision making. Three will be selected from each of the SGZs. Nominations for these will come from peers; effectively scientists will be nominated by scientists. They will hold long tenures of 25 years, there will be a complete lack of incentives to cheat, and they will be safeguarded from external influence. Comprising 11 women and 10 men, the balance of one will rotate over time.

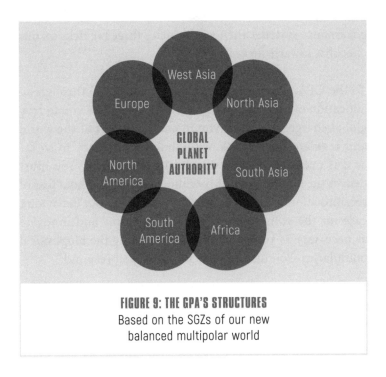

FIGURE 9: THE GPA'S STRUCTURES
Based on the SGZs of our new
balanced multipolar world

Reporting to the Biophysical Board will be the Operating Executive, which will have an operational centre in all SGZs. Besides more scientists, biologists, ecologists etc., the Operating Executive will consist of many of our brightest lawyers, taxation experts, logistics experts and strategists. These professionals will be responsible for ensuring that the biophysical boundaries are adhered to. They will build the global transaction fee system (next section), write the rules and operate the systems to protect the asset.

Once we have created the External Board, a Biophysical Board containing 21 of our best environmental scientists and an Operating Executive of excellence with best practice internal governance systems, then we can place three big ticks on our checklist toward success.

To the GPA, we will say: "We have created you through an allocation of our personal sovereignty. We allow you to levy global charges, we allow you to enter any part of the world and regulate the actions of the human populace in so far as it is necessary to deliver a healthy biosphere. You must deliver the function of utility without bias and without fear of recrimination. You must not kill. We will alter our behaviour, care for the vulnerable, adjust our economy and innovate as necessary as you go about implementing the biophysical boundaries. You just do your job, and we'll respond."

COMMANDEERING ASSETS: GLOBAL TRANSACTION FEES

Global transaction fees are an untapped resource capable of releasing huge financial power in the pursuit of biophysical integrity.

I believe that the GPA will raise and spend 6% of GDP p.a. for the first ten years of its existence, a 15-fold increase in spend compared to the current national environmental spend of 0.4% GDP. This would allow the GPA to spend approximately $0.75tn on each biophysical boundary and make a wealth transfer to the ultrapoor of 1.5% of GDP per year. After the initial ten years, the fees will be materially reduced as the GPA enters the long-term maintenance phase.

As all the charges are transaction based, you will pay them indirectly (i.e. you will not receive a bill) by way of higher prices as you go about your daily life, unless you are very rich and are subject to the direct property taxes I detail below. I urge you to see these global transaction fees as a utility bill. We are generally happy to pay our bills for our phones, electricity, water and sewerage, aren't we? In the main, they are great value for money as they are well-run utilities. We have to live on a healthy planet, so let's willingly let the GPA collect its utility fees – they are going to be the best value of the lot.

Go and argue, if you want, for reductions in your municipal, provincial and national government taxes. You could argue for removal of farm subsidies and fossil-fuel subsidies (together $800bn p.a. in 2018 (source: World Bank)) or for reduced national military spend if you think that is out of line.

Far better people than me will structure the global charges to be the most effective and equitable. However, I believe the fee structure could be like this (in my calculations I have assumed that the effective collection rate is 60% on all fees levied, in order to be conservative):

FLAT 3% GDP CHARGE WILL RAISE $1.4TN P.A.

An advantage of this charge is that it is uniform and broad. We would all pay it, on all **final** goods and services. As a guide, just think of your toothpaste being 3% more expensive. Not too bad, is it?

Another advantage of such a charge is that it is tied to GDP, so the amount of money collected would adjust up and down as GDP moves. The stronger and larger the global economy, the larger the amount of money there is available to spend in preserving and enhancing the biosphere. This tax is regressive in that it hurts the budgets of the poor more than the rich, but we can smirk when a rich person buys an expensive watch and pays 3% on that.

PROGRESSIVE TAXES WILL RAISE $1TN P.A.

'Progressive taxes' refers to taxes that are levied on the wealthy. If you have net assets of $1m or more, then you are in the wealthiest 0.8% of the world's population. We are talking about these folks.

Tobin taxes $0.25 tn. In the 1970s, economist James Tobin suggested a tax on currency trading. This has been widened in meaning to cover taxes on rapid transactions.

If the GPA taxes currency trades and derivative trades at 1 basis point (a basis point (bp) is one hundredth of 1%)

per trade and bonds at 5 bp per trade and shares at 10 bp, these taxes would raise $0.25tn.

Progressive Industry taxes $0.2tn. If the GPA decided to tax automotive industry sales at an additional 3% on top of the 3% global GDP charge, the luxury goods industry an additional 20%, and mining an additional 5% (each on turnover, so these would be inflationary consumption/production taxes), it would raise $0.2tn. At this point in time, given the task we must undertake, if you can afford a $2,000 handbag, you can afford a $2,400 handbag.

Property taxes $0.5tn. This is a straight wealth tax. This is the only one I propose that is not a transaction charge. Global property is estimated to be valued at $215tn (source: Savills). If we assume that all properties worth over $1m add up to 50% of this, then levying an annual 75 bp tax on these properties would yield $0.5tn p.a. Don't worry, rich people, it's only for ten years; you'll find the cash somewhere.

TAXING THE EXTERNALITIES WILL RAISE $2.3TN P.A.

These taxes would be levied against the industry that is polluting, so you and I will pay this if we continue to use the polluting product. Then as substitute products come along that don't incur such taxes as they don't hurt the biosphere, we can switch to them and avoid paying the charge.

Carbontax.org is a really useful website – thank you, guys. According to this website, at $125 per tonne, carbon taxes for the USA would be $440bn p.a. The USA accounts for 20% of the world's carbon dioxide emissions, so that would mean the tax would yield $1.3tn globally. By the way, for those of you reading this in the USA, it would add about $1.20 to your cost of a gallon of gas. Those of us living in Europe, Asia and elsewhere say: welcome to our world!

The level of carbon taxes will increase thereafter, but proceeds to the GPA will not rise materially, as substitution effects will kick in quickly. That is, we will rapidly use less carbon-based fuel. There will also be taxes on other 'bads' like nitrous oxide, methane and sulphur dioxide.

If petroplastics are taxed at 50%, then this would yield $0.18tn. The GPA may deem that our airline tickets need to be taxed further. A 30% tax would yield $0.2tn.

There will be no labour taxes, no interference with the tax systems of nation states, just separate GPA transaction-based charges that will be paid by us all, but the rich and the profligate will pay the most.

Armed with regulatory powers and $4.7tn of revenue each year (2018), the GPA will now be in a position to put in place the universal biophysical boundaries to ensure that we rebuild and then maintain a healthy biosphere for the ultra long term.

Just to repeat, whereas the structure of the GPA and the charges that could be collected are my own proposition, the following actions of the GPA and boundaries to be imposed are, in the main, a rough compilation of current recommendations by environmental experts.

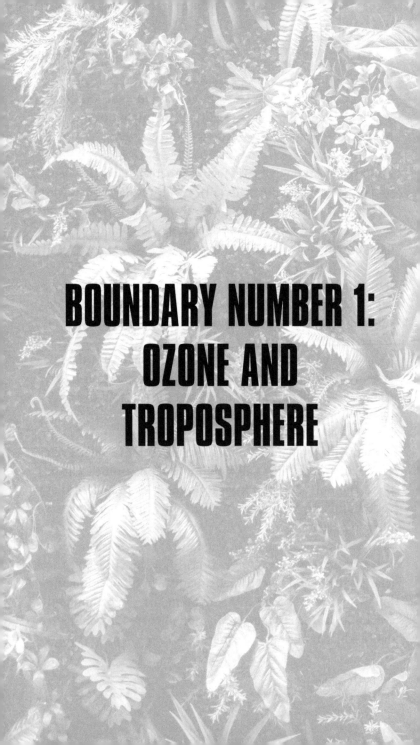

BOUNDARY NUMBER 1: OZONE AND TROPOSPHERE

GPA DIRECT SPEND $0.5TN P.A.

The GPA will put in place a global ozone monitoring system with rapid shutdown capability of any factory producing gases that could damage the ozone. For example, detection of any emissions of dichloromethane or dichloroethane (gases that can arise in some production of paint strippers and PVC) will result in the relevant factories being shut down, wherever they exist.

The GPA will have thousands of experts working on the decarbonizing solution every minute of the day.

The GPA will take our economy to zero CO_2 emissions in under ten years.

It will do this by applying an aggressive global carbon tax, well flagged to the market. There are currently some form of carbon taxes in 41 countries, which the OECD (2018) computes to be at €14 per tonne of CO_2. The current Swedish rate is €150 per tonne, although not applied to all electricity production, and Sweden's economy is the world's third most efficient. India, China and the USA, the three largest emitters, have negligible carbon taxes.

The GPA will start immediately at $125 per tonne of CO_2, and take this to $250 after four years and $500 per tonne of CO_2 after eight years (my figures alone).

By way of Adam Smith's invisible hand, this would activate the $6-$9 tn of annual capital investment (roughly 7-10% of global GDP) for the ten years that the IPCC and World Bank, respectively, believe is necessary for us to decarbonize.

We currently invest \$330bn p.a. on clean technologies, so it would be an increase of 20-30 times our current spend.

Our capitalist system would embark upon a huge upscaling of its research and investment into low-emitting cost-effective energy technologies. This global taxation of carbon dioxide emissions will spur growth in the production of solar energy without nitrogen trifluoride, drive down the cost of carbon capture and storage, fund a green energy revolution based on photosynthesis, and more.

Projects such as Europe's mega project of DESERTEC or Morocco's Noor 1, 2 and 3 will be initiated, battery plants will be expanded in the USA, and China's world-leading direct current technology will be in high demand. Australia may be the first country to have all its power made by way of solar thermal plants.

The GPA would also make important decisions on nuclear power and other technologies to come. But herein lies the beauty of the GPA: you and I will know that our experts at the GPA are making the best, most informed decisions on our behalf ... at all times. If nuclear is best, then so be it. If not, then so be it. And so it will go on, through the ages.

The GPA will also have two other weapons in its tool kit: regulatory powers that will enable the GPA to close the world's 1,000 dirtiest coal-fired power plants if it chooses to, and \$0.5tn p.a. of the GPA's own to spend for ten years. This would be spent wherever the GPA needs to rectify market inefficiencies or invest in research and development.

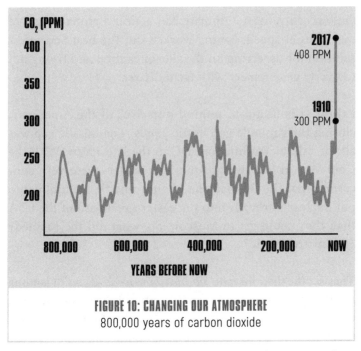

FIGURE 10: CHANGING OUR ATMOSPHERE
800,000 years of carbon dioxide

Source: Luthi et al (2008) (cdiac.ornl.gov) and Keeling et al (Scripps.ucsd.edu)

The GPA would immediately start work preparing a plan capable of removing CO_2 from the troposphere, with the target of reaching 300 ppm in 20-30 years. This closure of the carbon gap is something that we do not know how to do exactly. Scientists have provided a suite of actions which the GPA will most likely analyse: afforestation (planting trees on land where there were no trees previously), reforestation (planting where there were trees previously), the use of biochar, soil carbon sequestration, expansion of wetlands, expansion of bioenergy with carbon capture storage, accelerated weathering, direct air capture, ocean alkalinity enhancement and changing CO_2 to durable carbon (source: UN).

Only a centralized authority can action a global range of measures at speed, having worked out the best and safest solution for us, erring on the side of caution and trying not to overly geoengineer with techno-fixes.

I think it is useful to remind ourselves of the Americans closing their missile gap in the 1960s. The missile gap was all the talk of Washington, DC in the late 1950s. With the Cold War in full swing, American leaders were very concerned that the Russians had better missile technology to put nuclear warheads into the eastern seaboard of the USA than they could get to Moscow. So, what did the Kennedy administration do?

They set the then *we-have-no-idea-how-to-do-it goal* of landing the first person on the Moon. It captured the imagination of the American public, gained their approval for a portion of their taxes to go to the space programme, and focused the minds of their best engineers on one objective. The Americans set foot on the Moon in 1969, and they then had better missile technology than their Cold War foes.

But the relevant statistic is this: the average age of a NASA engineer in 1969 was 27 years old. These engineers had been first-year students at university when the declaration was made eight years previously. If the GPA declares we have to go to 280-320 ppm, you can bet that many of our global engineers and environmental scientists who who solve this problem will be at university when the declaration is made.

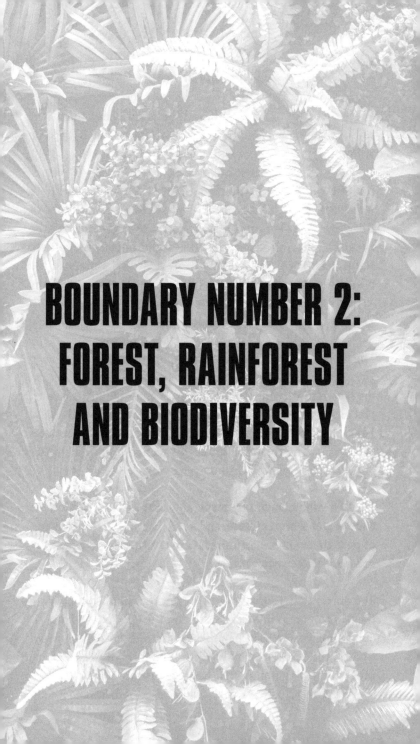

BOUNDARY NUMBER 2: FOREST, RAINFOREST AND BIODIVERSITY

GPA SPEND $1TN P.A.

The University of Maryland and the IUCN have identified 2bn ha of degraded land (not viable agricultural land) capable of holding forests. That is the size of South America.

The GPA will plant the full 2bn ha in ten years. Given there are almost eight billion of us, that would be one quarter of a hectare each. We can do that. The GPA will supersize the Great Green Wall project in Africa and boost the Chinese Natural Forest Conservation Project. It will supersize the Bonn Initiative, Initiative 20×20 in Latin America and the African Forest Landscape Restoration Initiative (AFR100). The majority of the trees would be planted by local groups.

The GPA will, for example, also employ the likes of Dr Susan Graham, who has helped build a drone system that can scan the land, identify ideal places to grow trees in hard-to-reach places, and then fire germinated seeds into the soil. As of 2017, she was hoping to use drones to plant one billion trees every year. The GPA could use this technology to plant five billion trees a year by drone.

The GPA will immediately place all of the world's rainforests under a global protectorate, and it will ban the terrible practice of peat fires. Global rainforest protectorates. The whole lot, in under three years.

The current GDP of all the countries that hold some part of the Amazonian rainforest is $3tn p.a. The GPA would have the ability, for example, to pay these countries 7% of their GDP, $210bn p.a., for ten years to secure the formation of a global protectorate. Assuming that tax receipts are 21% of their respective GDPs, then $210bn would be 33% of their tax receipts.

That goes a long way to improving the wealth of the nations involved and putting in place plans to get non-indigenous urban populations to move out of rainforests, in an action that is financially beneficial for them to do so.

The GPA would repeat this arrangement in the African and Asian rainforests. It will ensure that soya farmers, palm-oil companies, cattle farmers and loggers can no longer remove rainforest. They will have no chance.

The GPA will set and achieve the objective of bio-abundance, no loss of species. The GPA will achieve the objective of 40%+ of the world's landmass being in global parks that incur almost zero human pressure. This does not mean that there cannot be humans present but the global parks are maintained as ultra light zones. The world's biodiversity hotspots will be trebled in number, from 34 to 100, within one year.

Our ecologists (E.O. Wilson, among others, with work overlapping with hotspot analysis) have already prepared a list of the regions that must be safeguarded and enlarged first:

- Redwood forests of California, longleaf pine savanna of the American South

- The Madrean pine-oak woodlands of Mexico, Cuba and Hispaniola

- The whole of the Amazon river basin

- The Guiana shield and the Tepuis, the greater Manu region of Peru

- Paramos, Atlantic forests of South America, Cerrado, the Pantanal, the Galapagos

- Białowieża forest of Poland and Belarus, Lake Baikal in Russian Siberia

- The Christian Orthodox Church forests, Ethiopia, Socotra (islands)

- The Serengeti grassland ecosystem, Gorongosa National Park

- Mozambique, South Africa, forests of the Congo basin, the Atewa forest

- Ghana, Madagascar, Altai mountains

- Borneo, Western Ghats of India, Bhutan, Myanmar

- Scrubland of southwest Australia, Kimberley region, Gibber plains

- Papua New Guinea, New Caledonia, Polynesia including Hawaii

Let us imagine that the GPA's Biophysical Board decides that some of the areas of the world that need to be protected include the whole of South Africa, the massive Białowieża forest in Poland and the whole of Madagascar. They inform the Operating Executive, who then make the call:

"Hello, is that the President of Poland, South Africa, Madagascar?

"Yes? Well, it's the GPA here, we've got to talk... we need to take your national parks to global status immediately, increase them in size fivefold within five years and then work out a plan for taking your entire country to global protected area status, which we will pay for and subsequently employ a lot of your honest hardworking people."

Many more multi-nation migratory parks will be created. If the GPA decides it is necessary to create a migratory park through Kenya, Tanzania, Mozambique, Rwanda and Zimbabwe, then the GPA could offer $30bn p.a. for ten years for this to be implemented, primarily paying for a large conservation employment programme. The collective GDP of these countries is currently just $120bn p.a., so it would be an increase in their GDP of 25% p.a.

The Virunga National Park in the Democratic Republic of the Congo is home to the largest number of the famous silverback gorillas, 880 of them. But it is also home to armed guerrillas, and 85 oil licences have been issued within the national park. A Virunga global park will target 5,000 gorillas and zero oil licences. Enforced and paid for.

Some of the numbers are embarrassingly small. It is estimated that it would cost only $85m to save the tiger from extinction.

The GPA will set up a global invasive species command centre to upgrade and coordinate national systems to crack down on animal and plant trafficking.

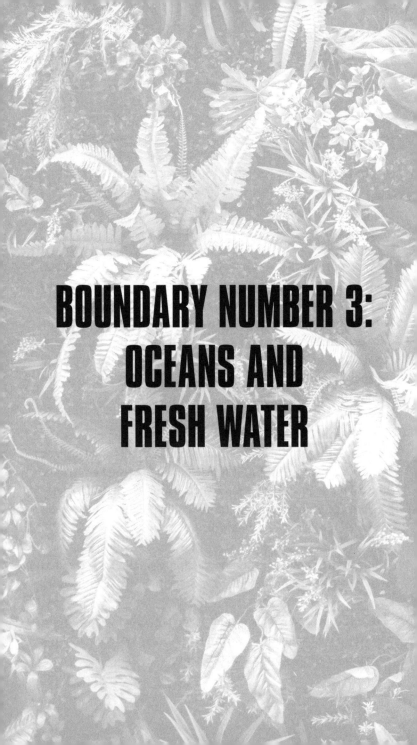

BOUNDARY NUMBER 3: OCEANS AND FRESH WATER

GPA SPEND $1TN P.A.

The oceans must be clean, cold, devoid of plastic and bio-abundant. **The GPA would rapidly take us from the current 2% no take to 40%-60% no take**.

The GPA will declare the Arctic and Antarctic to be globally protected areas, that is, there will be no national sovereignty in the region of the planet's ice caps. The high seas (outside national maritime boundaries) will all immediately fall to the GPA to manage. The deep sea trawlers of just five nations account for 80% of this destructive practice. The GPA will ban deep sea trawling immediately.

Every vessel in the world will be tagged using optical sensors, and ships diverted to new trade routes where necessary. At present our satellites turn off when they go over the oceans, in order to save energy. The GPA will ensure they are switched on all the time and the oceans properly monitored.

The GPA will set the global objective of a tenfold increase in large fish and marine mammals (whales, seals, polar bears etc.) in order to be back at the levels of 1900. Whaling will face a properly enforced global ban. The GPA will create mobile marine reserves, so when whales or particular groups of fish migrate, the reserves will move with them.

The GPA will remove the five plastic gyres of the world's oceans as fast as possible, hopefully before all the plastic breaks up. This is another thing we don't know how to do yet due to the size of the oceans, the waves, currents, wind, etc. Our GPA will dedicate vast resources to the problem.

One of the solutions that has captured the world's imagination is Boyan Slat's Ocean Cleanup autonomous floating capture ring. Founded in 2013, the Ocean Cleanup project now has 80 engineers working tirelessly to get their machine effective. Our GPA could, for example, take the project's workforce to 8,000 engineers.

CLOSE THE FRESH WATER GAP

The GPA will set the target of fresh water extraction at a maximum of 4,000 km^3. To produce 1 kg of bread takes 600 kg of water and to produce 1 kg of meat takes 13,000 kg of water (source: Institute of Mechanical Engineers). You might be surprised to learn these, like I was; there are not zeros in the wrong place.

Institutions such as the World Bank's Water Resources Group know we can achieve huge efficiencies via 'crop per drop'. We have field expertise in drip agriculture, subsurface drip agriculture and hydroponic agriculture. There are tremendous opportunities in Natural Sequence Farming, which reinstates the way water moves across the land surface, and in the use of dew and maritime fog. Crops will be planted that are more suited to the natural conditions rather than relying on our current method of diverting huge amounts of fresh water to plants that are not native.

The GPA will take the world to a 100% basin approach with 100% transboundary agreements on all rivers. Taking a basin approach (actually integrated river basin management in formal terms) places ecosystem functioning as the paramount goal. Only 40% of the world's 276 international river basins have any type of cooperative agreement. The GPA will declare that it has direct control over all river basins and set the rules that each individual nation state must follow to ensure the biophysical health of the particular river system.

The GPA will set the target of halving the amount of water removed or 'fragmented' out of the world's rivers, back to 30%, the pre-World War 2 level. The GPA would oversee a 30-year revolution in the way the human race control rivers. Any surplus dams will be removed immediately, as they have started to do in the USA. Every river will be set a maximum number of dams. Many may be set at zero. Now and in 3025, all rivers must flow from source to mouth.

The Aral sea, one of the most photographed 'inland sea disappearances' due to human bad management, will be restored.

The GPA will set up a specialist team to target the 13 aquifer hotspots and ensure that all of the world's 37 biggest aquifers are in good health. Among the tools at its disposal is the existing World Resources Institute's Aqueduct Water Risk Atlas.

Wetlands must be regrown and be healthy. They are very good at capturing CO_2 from the atmosphere. Besides all the mangroves, saltmarshes and other wetlands that need to be dramatically expanded, the GPA will double the aggregate size of protection of the world's seven major wetlands: Wasur, Camargue, Okavango, Pantanal, Kafue, Kakadu and Kerala.

BOUNDARY NUMBER 4: TOPSOIL, FIXED NITROGEN, PLASTICS AND LANDFILL

GPA DIRECT SPEND $0.5TN P.A.

The GPA will set the target of no topsoil erosion and achieve this within 20 years. Our global agricultural practices must be changed to preserve topsoil and use less water. Every farm must be changed.

We basically need a new way of farming, called agroecology. It involves much lighter tillage, sustainable planting schedules and natural yield increase. Essentially, we have to partially un-simplify our agriculture away from the almost complete monoculture of today to mimic nature's biological communities, whose diversity provides strength. The UN believes that yields can be doubled on current agricultural lands via mass agroecology roll-out.

We know, for example, that fields where Asian rice is grown benefit greatly from crop diversification. Growing flowering crops in strips beside rice fields has been demonstrated to reduce pests effectively because the flower nectar attracts and supports parasitoids and predators. By doing this one action, insecticide spraying is reduced by 70% and yields are increased by 5% (source: Geoff Gurr and Zhongxian Lu. 2016).

The GPA will limit the total area used for growing crops to ensure it does not surpass its boundary, and start to put carbon put back into the soil globally, a method the Australians have mastered in their carbon farming.

The GPA might decide that we need to re-implement the mass livestock herding of the Holocene era.

The GPA will reduce our use of fixed nitrogen by 75% in 30 years by way of regulation. This is going to be a very difficult and slow task, as we rely so heavily on this element for our food production. The GPA will firstly aim to minimize run-off of existing nitrogen-based fertilizers from farms, before overseeing genetic engineering of plants that are able to take nitrogen straight from the atmosphere like legumes do. Waste paper and CO_2 captured fertilizers will be part of a suite of replacements for some nitrogen-based fertilizers, including human excrement from cities.

The USA Environmental Protection Agency's nitrogen oxides budget trading programme, which has proven to be very effective, will be copied and rolled out worldwide.

The GPA will set up a global chemical pollution monitoring team with automatic shutdown authority. Of the 75,000 man-made chemical compounds, we have only properly tested 5,000 for environmental effects. The GPA will allocate the resources to take us to 100% tested.

The GPA will reduce our petroplastic use by 80% in 20 years. The GPA will use hypothecated taxes to effect the change. Hypothecated taxes are those where the beneficiary is clearly defined, such as when cars have to pay a fee to pass across a bridge and that money is used to maintain the bridge.

By aggressively taxing petroplastics with hypothecated taxes globally, the money will flow into rapidly growing its substitutes: the bioplastic and sustainable packaging industry will grow at an extraordinary rate to replace the current petroplastic output.

There will also be specific taxes on any plastic reaching landfill, the half-life will be published on every piece of plastic, and there will be fast track introductions of new technologies such as plasma torch technology and pyrolysis, which is the thermal decomposition of materials at elevated temperatures in an inert atmosphere. Essential services that rely on many plastic instruments, such as an operating theatre or an ambulance, will have specific plans to ensure no interruption of services.

The GPA will also set the goal of **zero** landfill in 30 years. It will achieve this by setting very punitive taxes and building global monitoring systems, which will lead to massive investment in recycling with new separation techniques.

INTERNAL BOUNDARY: FEW ULTRAPOOR

Should the Operating Executive of the GPA decide that it is the right action necessary to gain biophysical integrity, the GPA will oversee the biggest wealth transfer to our ultrapoor ever undertaken. The GPA would have the ability to transfer 1.5% of GDP p.a. for the first ten years to the poorest citizens on Earth, which is five times the current direct aid contributions (DAC) of 0.32%.

Study after study, field example after field example, show that direct money transfer is the most effective at liberating poor individuals. They know best as to what they should spend their money on, dividing it between education, energy security, food security, birth control and internet access/ mobile phones.

We now have the technology to make safe and secure direct money transfers without encountering any intermediaries and their corruption.

CONCLUSION

This is what success looks like.

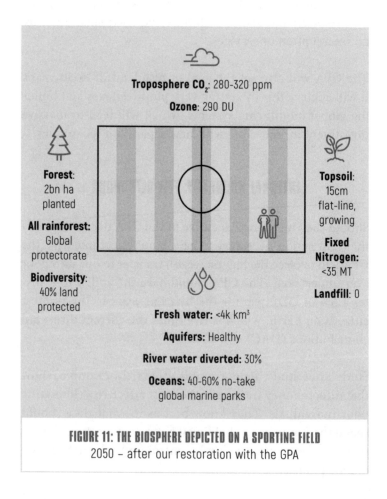

Troposphere CO$_2$: 280-320 ppm

Ozone: 290 DU

Forest:
2bn ha
planted

All rainforest:
Global
protectorate

Biodiversity:
40% land
protected

Topsoil:
15cm
flat-line,
growing

**Fixed
Nitrogen:**
<35 MT

Landfill: 0

Fresh water: <4k km^3

Aquifers: Healthy

River water diverted: 30%

Oceans: 40-60% no-take
global marine parks

FIGURE 11: THE BIOSPHERE DEPICTED ON A SPORTING FIELD
2050 – after our restoration with the GPA

We can accept nothing less.

Only a bespoke supranational authority can act at this speed and with this amount of force. Our current system will never achieve it. It is time to replace the part-time amateurs with full-time professionals.

Besides focussing on the granular, having a central authority capable of overseeing a holistic approach is also key. I have not attempted to convey the work that needs to be done in the many instances where solutions involve several of Earth's systems. For example, to protect corals, it is calculated (e.g. Lynas) that the boundaries of climate, nitrogen, biodiversity and ocean acidification must all be solved in order for us to restore coral reefs.

Executing its mandate, the GPA will ensure that humanity does no further harm to the biosphere. The GPA will use sufficient force of both regulation and monetary spend, combined with intelligence, skill, expertise and complete focus on its task to succeed.

I suspect that you might well have calculated that it is possible that the world's total investment spend on protecting the biosphere and reducing our industrial metabolism could be as much as 25% of GDP p.a. for many of the first ten years of the GPA's existence. This would comprise the 6% GDP direct spend by the GPA, 10% of GDP being spent on capital expenditure in response to CO_2 taxation, plus potentially a further 10% of our GDP being invested on substitutes in response to the taxation and regulation of other externalities.

If this indeed turns out to be the case, I believe that this would amount to our imposition of Newton's third law of an equal and opposite force. It would be the appropriate human response to existential threat and the proper shouldering of our greatest intergenerational responsibility. It would build our legacy.

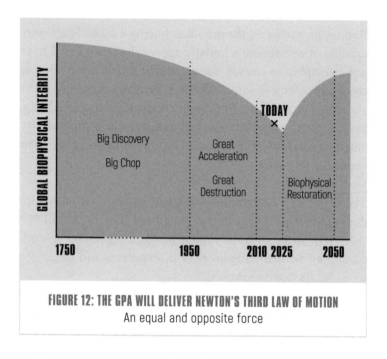

FIGURE 12: THE GPA WILL DELIVER NEWTON'S THIRD LAW OF MOTION
An equal and opposite force

We need the machinery that can deliver us global biophysical boundaries now. In order to rapidly reduce the risk we face and to set us on the path toward long term biophysical integrity, we need a supranational authority, we need the right global lever, nothing else will do.

OUR RESPONSE

In response to the imposition of these boundaries, there will be a technology, education, engineering, employment and use of waste revolution.

Because it is bigger than ourselves, because it has intergenerational consequences and will involve a lot of hard work, sacrifice and change, it will be bonding and purposeful. We will move from being at risk at the end of the toggle of the yo-yo to being on the next biophysical sphere with clarity, structure and safety.

Our biophysical boundaries will not be constraining, they will be more liberating than we can imagine. We will quickly innovate, change and grow.

With a GPA in place we will be able to execute two positive jobs: to preserve and restore the biosphere, while continuing to improve our human condition. It will no longer be one at the cost of the other as it has been for the last 270 years.

Our responses down the line will be superb. From the individual to the family, from the workplace to the individual nation. Our brilliantly diverse human race has so many strengths and qualities to offer this undertaking, especially our ability to cooperate, to use our imagination and to innovate. It will be hard but hugely rewarding work.

Capitalism's response will be rapid, efficient and effective. Capitalism is waiting for market prices and physical boundaries to be set. It is young and strong and wanting to be tested of its capabilities in a more materially finite operating system.

We are going to transition into being a net positive force and evolve out of being a pioneer species to a more mature species that is self-renewing. Effectively we are going to become ants.

"I don't want to become an ant!" I hear you cry!

But don't worry, the humble ant's biomass is bigger than ours, yet everything that ants build, such as cities and farms (they grow mould in purpose-built underground farms), is completely renewable and biodegradable. Everything.

Our industrial methods and materials will change from being depletive to intentionally augmentative. Industrial metabolism will move down and natural metabolism will move up.

We will have a design revolution as we realize that our future involves no waste. We'll make televisions where every component is renewable; our engineers are going to work out how to recycle the mix of polymers and all the other 4,400 chemicals in a TV set.

We will plant massive bamboo plantations, grow algae on the edges of the desert in tanks that are biodegradable, and every part of a car will be 100% renewable. Automotive manufacturers will want all of their cars back to recycle.

There will be massive seaweed farms, sequestering CO_2. We will build huge mushroom farms grown on ground-coffee waste, and there will be a transformative expansion in fertilizers made from CO_2 and waste products. A lot of shorter supply lines will be created to augment our global ones, which will drive growth in local employment, and there will be extraordinary waste-to-energy advancement.

Synthetic paper, already invented, will become commonplace, where you can wash off the ink and reuse the page, meaning we don't need to cut down as many trees. Buildings will be made with carbon negative cements, and the buildings themselves will produce more energy than they consume, also consuming their own waste water. Leather will be tanned without chromium, immediately.

There will be a water and soil preservation revolution. For example, in Palestine there are schemes that take manure-ridden water from when farmers wash down their animals, put it in tanks and use the methane released to cook with; they then plant Dutchweed in the tanks. The Dutchweed grows and cleans the water and it is itself is a good foodstuff for chickens and ducks and the resultant clean water is used for irrigation.

Nanotechnology and robotics will drive our new intensive, rather than extensive, economy. Synthetic biology will ensure transformational microbe-based increases in food and energy. A significant portion of our meat will be grown in laboratories. We may fish using bubbles instead of nets, as dolphins have done for a million years.

Good industrial design will not even feel the biophysical boundaries, because it will be so far within the boundaries of the playing field.

To the young people reading this, you might be concerned by the recent warnings by some 'experts' of mass unemployment from the widespread introduction of artificial intelligence.

Instead, with a GPA in place, we will be protecting and restoring a biosphere while changing nearly every single aspect of how we build our human-made capital. You're going to be super busy!

We are going to embrace the challenge, be bonded in our collective ambition and bear witness to our progress.

Of course, the GPA will not be a panacea for all of humankind's ills. It will not oversee the removal of our 15,000 nuclear warheads, solve the drugs crisis, close the wealth gap, or stop spikes in the rate of teenage female genital mutilation in war zones. The GPA will not address obesity, mental health issues, remove guns, or stop young men being rounded up and murdered in acts of genocide.

This will not be nirvana. It will not be the end of history as we know it, a proposition put forward by Francis Fukuyama at the end of the Cold War. We humans will still find it in ourselves, well, to be human. But by putting in place a GPA, we will give the planet the chance to sustain us and all other life forms – and we can go on trying to improve the human condition.

FOUR IMMEDIATE QUESTIONS

You haven't addressed the issue of population, so will it work?

Yes, it will work.

Surely, goes the argument, if the size of our population + capitalism + consumerism = today's huge force of destruction, then unless we reduce our population, replace capitalism and materially subdue consumerism then we will be ineffective in reducing our overall impact.

The answer to that is no. It comes from an existing governance point of view, i.e. remove the inputs as there is no effective governance protecting the biosphere. As soon as we put the GPA in place, then it will have sufficient power to move us to biophysical security.

The solution is not in removing the source itself, but altering the direction of the force being exerted with a bloody big kick and placing humanity on the playing field, whatever size and shape we are in. Think of the potency of taxing cigarettes and the resultant incidence of smoking over the last 30 years.

Let's cast our minds forward and imagine a scenario in 200 years' time. The GPA is tending to its job of maintaining biophysical integrity, but now socialism is dominant, our population size has fallen to five billion due to a moderate pandemic or reduced birth rates, and the number of nation states has halved due to the formation of the large single nations of West Africa, Europe and the United Middle Asia. The populace is consuming intelligence-enhancing drugs and living to an average 150 years of age.

Should the GPA intervene to boost population size and reduce longevity, or change the operating system of socialism
or the consumption of drugs? No. That would be outside its
mandate. The GPA will have sufficient power to do its job
without these interventions and will continue to be guided
by its best practice external and internal governance systems.

Will we starve?

It will be argued that to cut our fixed nitrogen production
(used as fertilizer) and potassium use while rapidly replacing
CO_2 producing machinery is folly in the extreme, because
hundreds of millions of people will be put at risk of starvation
due to the reversal of the green revolution of last century.

We will not starve. There are multiple reasons for this.

First, with the knowledge that 66% of our calorific input is from
maize, rice, wheat and soya, the GPA will have a dedicated team
who transition us to ever more environmentally friendly forms
of fertilizer while maintaining total calorific output. The GPA
will sensibly contain and reduce our fixed nitrogen use over
time, and national governments will work with industry to
strengthen key transport links of essential food items.

Second, the UN has estimated that yields can double with
agroecology on existing farmland, and that programme will
get underway at speed.

Third, excess capacity. A global economy that is capable of
making plastic luminous wands for Halloween that are often
broken within five minutes of their compulsive purchase has
excess capacity relative to basic needs.

Fourth, we are already overweight. According to the World Health Organization, in 2016 more than 1.9 billion adults (35% of adults, 18 years and older) were overweight. Of these over 650 million were obese. Most of the world's population now live in countries where being overweight and obese kills more people than being underweight.

To supply the global demand for essential food in the short term will be a matter of diverting investment away from surplus consumerism and bad food. No bad thing for a little while, I would have thought.

Last, we've heard the starvation argument many times before in opposition to governance change. For example, we heard it every single day for 30 years prior to abolition of Slavery Act in the British Empire in 1833. "You cannot free 900,000 Caribbean slaves – the British Empire will starve", went the argument. It was complete self-serving rubbish; no one starved in 1834 or the subsequent years due to the abolition of slavery.

Will there be an economic depression or material GDP contraction?

Quite probably, but the growth thereafter will be faster and more sustainable. It is very important for us go into this with our eyes wide open. There could well be an economic shock; however, this is a burden we will anticipate and bear with determination.

It is important for us to remember the difference between the worldwide depression of 1929-32, which was due to bad policies, excess speculation and the stock-market crash, and the human suffering it caused, and the economic depressions witnessed by countries during the early years of World War 2, which were due to the amount of capital diverted into financing their respective war efforts. These depressions were technical ones, as the diverted capital was assigned a less commercial value of output, but people were still fully employed, having moved into new jobs, and did not starve.

Knowing a disruption is likely, we can plan for it. An economic depression, if we have one at all, will be relatively short. We will be employed and redeployed, and with the market being told the prices of externalities in advance by the GPA, it will act quickly to press through substitution measures, reducing the duration of transition.

Thereafter it is very likely that there will be a long-sustained period of rapid economic growth. This is because the imposition of biophysical boundaries will accelerate growth as we rebuild our human-made capital stock at a new, much lower level of industrial metabolism. I believe that it will be similar to the economic activity we witnessed after World War 2, at the least. In fact it could well be faster, as per the industrial revolutions.

The growth of the OECD nations was 4.9% p.a. in the 1950s, 4.7% p.a. in the 1960s, and 3% p.a. in the 1970s, as we rebuilt, undertook the green revolution and the size of our global population grew.

Let's run some numbers because for all of us there is an element of "It's the economy, stupid" (coined by political strategist James Carville in 1991), even though we all know deep down that the pursuit of biophysical integrity is now immediate and preconditional.

Starting in 2025 with global GDP of $100tn p.a. ($87tn in 2018), GDP is forecast to grow by 2.4% p.a. to 2050 according to the OECD. If we extend this out five more years, our economy will be **$229tn** in size in 2055, assuming that a deteriorating biosphere can support it, which is unlikely.

Now let's assume that we form the GPA in 2025. With our GPA quickly imposing biophysical boundaries, we may well go into short-term economic shock due to regulatory imposition and massive capital diversion. If global GDP falls a total of 20% over the course of five years, the same amount that GDP fell in the USA in 1929-32, and then grows for the subsequent three decades at 4.9%, 4.7% and 3% respectively, our GDP in 2055 will be **$274tn**, bigger than continuing down the disastrous path without a GPA.

Are we too late?

If we rapidly reduce our impact and restore what we can using a GPA, is the damage done already too great? Has our human mosquito annoyed Mother Nature too much? Some think so:

"But the planet will not settle into a state that looks like the Holocene – the 10,000-year epoch of mild and constant climate that permitted civilization to flourish. It has been diverted onto a different trajectory. The idea of the Anthropocene was conceived by Earth System scientists to capture the very recent rupture in Earth history arising from the impact of human activity on the Earth System as a whole."

Clive Hamilton

But that sobering thought does not change the path that we must take. We must reduce our impact rapidly and then see. Unless we impose boundaries, we'll never know the answer, as we'll be in biophysical hell anyway.

Let's not ever, **ever** underestimate our ability to respond. I believe we can have a healthy bio-abundant planet and a bigger economy whose industrial metabolism is hugely reduced. We can undertake an action that is urgent now. We can put in place a governance structure that can take humanity forward, which our future generations will take for granted.

OK then, let's get on and actually create a GPA.

CONNECTED AND NO PERMISSION REQUIRED

In 2000, the number of us connected to each other by way of the internet was 0.75 billion and just 18 years later, at the end of 2018, it was four billion. Although the overall penetration rate is slowing (in 2017 the number of new users added was 0.25 billion), the number of internet users is on its way to five billion in 2022 (source: Cisco VNI) as African, Indian and Middle Eastern penetration rates grow quickly, catching up with or even surpassing the rest of the world.

We now have the ability to form a quorum as we have developed into a huge, closely connected global citizenship.

If we are going into a space where we haven't been before, that of global governance, then we have to go well past anything that exists at the national level. It has to be an emphatic statement and, as I detailed in the proposal chapter, I believe 1.5 billion is the minimum number. Remember that the previous number involved in an action of global self-determination was zero, i.e. there hasn't ever been one.

Our action to create the GPA **will not** be a vote imposing universal national democracy. How a country runs its affairs is up to the populace of that country, whether it be centralist, partial democracy, non-secular leadership, full democracy, secular or monarchist, etc.

It **will not** be a vote creating one global government where all nation-state sovereignty is ceded into one singular global state and we do not have countries any more.

It **will** be a vote that will create one vital supranational authority that has one specific job to do and has power over all nation states and all human organizations, and you and me, in order to execute its specific mandate.

Whose permission do we need to seek?

Respectfully, do we ask the UN, a body that is not voted into power, has a tiny annual budget of $10bn and no jurisdiction over the nation state? Do we ask the leaders of China, India, Indonesia, USA or Russia, the leaders, respectively, of our largest centralist state, our three largest democracies in the world and the largest country in the world by landmass? Not only will all leaders come and go over time (some slower than others, admittedly), but far more importantly, the offices that they hold cannot deliver us the function of utility which is now preconditional to humanity's future.

No, there is no one to ask except ourselves. By ceding part of our own personal sovereignty in sufficient numbers, we will be forcing the nation state system to cede part of their sovereignty to the higher bespoke authority. I hate to say it, but it will change the world.

It is this moral right of the quorum, gained by acting in the best interests of the whole and our future generations, that grants the power of enforcement to the GPA. This opportunity did not exist at the end of World War 1, nor did it exist at the end of World War 2, but it exists now because we are connected at the global level for the first time. Guns are not required, it is numbers of people … allied to a singular objective.

OUR HUGE NUMBERS, OUR INCREDIBLE STRENGTH

To gain 1.5 billion votes is a lot. Can we achieve this? An emphatic yes.

First, we are going to let anyone who is 13 years and older vote, so that nearly all the connected five billion of us will be eligible.

The reason is that it would be more analogous to include the teenagers than to exclude them. Throughout human history at points of revolutionary change, they have always been active participants: in Paris at the end of Louis XVI's reign, with Sun Yat-sen, with Gandhi. When Sun Yat-sen founded the Tongmenghui party while in exile in Japan in 1905, 90% of its founding members were 16-25 years old.

Today's teenagers are very smart, very connected and they understand the problem, often in more detail than their parents or grandparents, as they are learning all about it in school. And if someone is 15 at the time of the vote and I am 55, then they have 40 years more than me to live on a planet that is currently going to biophysical hell, so we **must** let them vote.

Second, within our connected five billion, we have four super groups, overlapping in an extraordinarily powerful Venn diagram. Our four super groups are the Global Generation, Greater Asia, Women and Men.

TWO BILLION CONNECTED YOUNG PEOPLE: THE FIRST EVER GLOBAL GENERATION, AGED 13-30 YEARS

You are our first ever Global Generation. You are hyper-connected. The digital age started around 2002, when the volume of our data stored in digital form surpassed that held in analogue form. So, if you are 13 years old, you have lived your whole life in the digital age. If you are 30 years old, you have lived your entire adult life in the digital age.

You make up the vast majority of the three billion social media users. You take a song ("Gangnam Style") by a South Korean pop artist called Psy, which is about the behavioural habits of the rich youth in Gangnam, an area in Seoul, and make it the first ever YouTube pop video to have a billion hits, the first to have two billion hits, and, as I write, it still remains in second place, with 3.1 billion hits.

If you've been on a flight, you get off in Hong Kong, New York, Buenos Aries, Mumbai or Sydney, get wifi and a coffee and then ask what country you are in. You can game, send a picture message on social media, watch Manchester United, Barcelona, Virat Kohli, a music clip and Steph Curry almost all at the same time.

Besides your global hyperconnectivity and the fact that you are smart, you have another trait that humanity needs at this time: your ability to see the world anew.

And anew it has to be.

It could also be argued that you are a generation of the secular state. But that does not mean in any way that you are devoid of spirituality, and your creating the GPA will be both an expression, not the only one of course, of your spirituality and an expression of your solidarity as a generation.

As the first ever global generation, your numbers mean that you can put the GPA in place by yourselves, an opportunity that will never have been presented to any group of people of your age in human history. You have a global problem and you can achieve the global solution – today.

TWO-AND-A-HALF BILLION CONNECTED PEOPLE IN GREATER ASIA

If you look up a list of the countries with the greatest numbers of internet users, our populace in Greater Asia (Mumbai to Tokyo, Beijing to Jakarta) dominate.

In our modern world, we live with its noise and distractions, short-termism and information overload. Yet we stand at a particular pivotal point in human history where we need unity and quiet reflection and we have a vital decision to make. Your pursuit of wisdom and civilization is by far the oldest in the world. We need your wisdom now.

For example, Confucius teaches us that the root to goodness is wisdom, that we should love the ritual and become *junzi*, which in Chinese philosophy is a person whose humane conduct makes him/her a moral exemplar. One of the greatest thinkers of all time instructs us to be trustworthy, humble and, I repeat, to love the ritual. At this vital time we need you to put in place a GPA so that we all have a global ritual of caring for the planet.

Ahimsa is one of the ideals of Hinduism. It means that we should avoid harming any living thing, and also avoid the desire to harm any living thing. We now know that the biosphere is a global living entity. We need the Indus population to vote into existence a GPA so that all of us refrain from injuring our most valuable asset.

The wheel turns again, back to the East for leadership.

TWO-AND-A-HALF BILLION CONNECTED WOMEN
AND TWO-AND-A-HALF BILLION CONNECTED MEN

I believe that if humanity's governance and progress was symbolically a fish, we have been swimming for far too long just flapping our tail to one male-dominated side. Eventually the fish just swims around and around in a circle and gets nowhere.

The creation of the GPA is an opportunity for us all to progress, to correct our direction of travel and build on the relatively recent foundations of suffrage with an action reinforcing universal female voting power. It is time for **all of us** to build an authority that upholds the sanctity of all the diverse life on this planet and does so with the greatest intergenerational care.

Each super group could vote the GPA into existence alone, due to their numbers, but together we are overwhelming.

Together we are an omnipotent force.

Together we are an omnipotent force for necessary change by way of global self-determination.

*"We **are** the people we have been waiting for."*

Amory Lovins

Yes Mr Lovins, you are right.

THE STEPS TO
THE VOTE

STEP ONE: OUR FIRST JOB IS TO ACHIEVE 150 MILLION OF US DECLARING THAT WE WOULD VOTE

This first step is your job and mine. In the 17th century village, you could walk next door, stepping carefully over the rubbish, speak to your neighbour about the crisis and start the process toward putting in place the new authority. That might have taken about a minute.

In the 21st century village, you can now go online, speak to your global neighbour about the crisis and start the process toward putting in place the new authority. That now takes about a minute.

It is key that we use a simple and very disciplined approach to our spreading of the idea and building our movement.

We must not criticize others or particular industries, governments or government officials.

The fault is ours for not having built the right governance system, so who is there to criticize? And we are going **past** existing structures in a clear, respectful and peaceful manner, not bludgeoning our way **through** with sharp elbows while hurling abuse at people.

Once 150 million of us have declared that we will vote, I believe that the momentum and the capital behind us will accelerate.

STEP TWO: THE VOTE EXECUTION TEAM

Holding a robust, verifiable global vote will cost quite a lot of money. For a group of philanthropists to write perhaps a $5bn cheque, they would need to see 150 million of us already signalling our willingness to vote. This is because it is reasonable to back a movement with a large amount of capital if success requires only increasing the current voter base tenfold, not having to increase it 100-fold or 1,000-fold.

And what an opportunity for the philanthropists: to enable the first action of global self-determination to take place by facilitating the will of the global citizenship.

And beyond that, if they want to think in monetary terms, they will get operating leverage on their capital countless times over. For example, if the GPA spends $4tn p.a.for ten years, that would be a multiplier effect of 8,000x on the philanthropist's investment, before the other investment effects are included.

When the cheque has been written, a different team will take over the vote process, a Vote Execution Team (VET). The VET will organize a global advertising campaign, optimize social media and assist in the formation of the Biophysical Board and selection of the Operating Executive.

The VET will ensure voters have verifiable individual voter codes gained by way of fingerprint or face recognition technology or other biometric testing, and put in place plans to ensure that our poorer citizens, who may have limited or no access to the internet, get the chance to vote. Using Blockchain or the like, the VET will enhance the prevailing technology to securely hold the vote online.

A specialist team from the VET will work with the Operating Executive to fine-tune the exact details of the vote in order for it to be best practice. The VET will most likely implement an opt-in-only vote. That is, a citizen will abstain by not voting.

The Operating Executive will have another important task to complete just before we vote. It will need to interface with the lead institutions that we are about to go past and forge preliminary lines of communication on behalf of the GPA.

STEP THREE: IT IS IMPORTANT FOR US TO VISUALIZE THE DAY OF THE VOTE

We have just voted, predominantly on our phones, and 1.5 billion, 2 billion or 2.5 billion of us have said yes. We are not presenting that as a petition to some existing body, because we will have just entered the global governance void **by ourselves** and brought the whole of humanity along with us.

Instead, there will be an extraordinary moment. History teaches us that it always happens – everyone will just stare blankly for a few seconds, recognizing that self-determination has won the day.

All the work, the talks, the capital spend, the planning, the threats against us, the criticism, the logistics, the advertising, the fieldwork to ensure biometric identity validation, the anti-hacking systems, the expansion of computing power on the net, and then the moment arrives.

And nothing happens. Everything just goes quiet for a tiny fraction of time, a minuscule vacuum. And then after literally one minute's pause, it will end with what I can only describe as a 'thunderous administrative cacophony' and people rushing in all directions to put new systems in place under the new governance structure.

The Biophysical Board, who have already met numerous times in shadow form, will hold their first formal session later that day. The Operating Executive, who have split themselves into two teams over the previous year, will commence their transition strategy. The pre-vote team, who had focused on legal, logistical and governance construction, will be replaced by the Stage One team, who have been meticulously preparing their plans for the first 90 days of the GPA's existence.

They will execute: the release of pre-arranged International Monetary Fund and World Bank funds and the immediate issue of a bond (already underwritten) to get funds into the GPA accounts and commence the set-up of the GPA centres in all seven geopolitical zones. They will create 50 new biodiversity hotspots, which have been preselected by the Biophysical Board, in just six months. They will dispatch taxation experts to administer the $125 per tonne carbon tax across all nations. They will introduce pre-arranged Tobin taxes on the world's financial exchanges, convene an emergency meeting of all nations whose lands contain rainforests, and there will be an immediate review of all underwater mining licences and an immediate ban on all deep sea trawling.

It will be one of the most extraordinary days of our lives. It will be the inflection point.

THE ARGUMENTS
WE'LL HEAR

As you and I start this journey, showing our willingness to vote and binding together in one movement, it is vital to anticipate the opposition and the arguments beyond the ones I have already discussed: it won't work as the GPA is not reducing population size, we'll starve and there will be an economic depression. Other arguments will include:

"This is a Utopian dream. Thomas More wrote of Utopia in 1516, and now you lot come along 500 years later. It was a pipe dream then and it is a pipe dream now."

If proposing an alternative path, an alternative horizon, that is rational and preserves the life system of this planet is a utopian pursuit, then call it what you want. We readily admit that the GPA will not be a panacea for all of human ills.

But we can say this with 100% conviction: it has the best chance of preserving the biophysical integrity of this planet, our one and only home, now and for all the generations to come, because it releases commensurate power and it fits the nature of the asset. Global for global.

Our first move into global governance will be very specific. The authority that we set up will have a clearly defined mandate pertaining only to the biosphere. It will not do all things for all people, instead the GPA will be razor sharp in the execution of its mandate in order to maximize the chances of success. We recognize that if there is any ambiguity or mission creep, then it won't work. The GPA will **only** act to secure biophysical integrity. Its external and internal governance structures will ensure this is the case.

"It will cost me too much: it's another tax on me, the middle-class person, and it will be inflationary."

When something is necessary, we've got to be noble, make sacrifices, do it and bear the personal cost. This action fulfils our greatest intergenerational obligation. So, let's front up.

Barring the property wealth tax, the transaction-based fees I have detailed are a balanced mix of regressive, flat, progressive, industry, manufacturing and consumption taxes; they will not be too onerous on the middle class, and you can avoid many of them by changing your lifestyle, if you wish.

It is true that many of the taxes I have detailed are likely to be inflationary in the first instance, but adherence to global biophysical boundaries will drive investment and huge efficiencies over time, which ultimately will be more deflationary.

"National sovereignty is sacred – they'll never agree."

We created nation states and now we must go past them. We will have to be completely focused and unwavering in order to do so, but we are going to because the blame for this dire situation lies with us, the solution is a global one, and we must rectify our error.

At the end of the day I think most of our national governments will be relieved. They have so many other things to concern themselves with, from obesity to drugs to healthcare and ballooning pension deficits. If the long-term health of the biosphere is ultimately the responsibility of a specialized team and the national governments are getting clear instructions as to what to do, then great, that's one less thing to worry about.

If we have to go, we have to go. Our nation-state governments can fall into line.

> *"This is just another grab by rich capitalists to be in control, and it would release biophysical tyrants, dictating globally with no accountability."*

On the contrary, this is about self-determination of the people of the world to put in place an authority we want to create. It is quite likely that a large amount of philanthropic capital will be required to hold the vote, and this money will probably come from a rich capitalist, but that is facilitation of our will, not leadership by them.

With respect to releasing biophysical tyrants, the GPA's Biophysical Board will be made up of 21 scientists, and besides the excellent governance systems that will be in place, the seriousness of their work and their intergenerational responsibility will be their moral guide. There will be a complete lack of incentives to cheat, tenures will rotate while maintaining continuity, and the use of technology improves transparency. That is, we'll be able to watch them and hold them accountable.

> *"Think of the world's poor; isn't this just another burden for them to bear?"*

No. A healthy bio-abundant planet benefits the poor the most. In the short term, the rich have resources they can use to mitigate some of the effects they encounter by way of biophysical degradation, whereas the poor do not.

It is true that a straight global 3% GDP fee is regressive, hurting the budgets of the poor more, and that electricity prices are very likely to go up in the short term. But opposing this are a greater number of progressive taxes that are borne by the rich and the profligate the most.

With the GPA in place, our expanding economy will have a larger number of short supply lines, more labour-intensive farming and see a surge in local employment as conservation schemes are supersized, all of which will benefit the poor.

Further, as part of fulfilling its obligations to deliver a healthy biosphere, the GPA will most likely effect the biggest wealth transfer to the ultra-poor that has ever occurred.

"There are too many variables here for me to work out, so it must be impossible."

My favourite argument for not moving to global governance of the biosphere is that there are too many variables to work out. Your mind does race, doesn't it, when considering just some of the variables:

- On the day of the vote, will the internet be shut down by nation states?

- Should the vote take place and the quorum of 1.5 billion be reached, will the Biophysical Board just walk into the UN even though we've gone above and beyond it?

- How do you build a global taxation system exactly and accurately locate buildings valued at $1m or more in order to tax their owners?

- How do you decide who is indigenous in the Amazon and who is not?

- How is the GPA going to avoid losing billions of dollars in corruption to rogue characters who are experts at taking their rotten share, normally at gunpoint?

- Despite the requirement to not kill, will the GPA have its own navy, its own small army and satellites ... will specialist soldiers drop into the middle of Vietnam, Uruguay, France and Ethiopia to expand the forest cover and declare global park status if the countries don't cooperate?

- Which dams are deemed essential for energy generation?

- How costly will the new, perhaps lengthened, sea trade routes be to the world economy?

- Etc., etc., etc. times a million.

This concept is called the fog of war. It states that the single human brain cannot compute all the variables of a multi-front global war. Individually, we cannot do it. Please read the following 50 times: **we, as individuals, cannot compute all the variables.**

We must get over this objection, because that does not mean that success is not possible. The right people will be working at the GPA to do the right job, in the right way. They will have our permission to act, the sharpest minds, the resources, the best team work and singular purpose. And existing organizational structures will have tasks assigned to them, all working to implement the change directed by the GPA.

"It's not the right time."

This argument being put forward is as certain as death and taxes.

Our pursuit of the GPA will always be 'at the wrong time' according to someone; in fact, it will most likely be argued it's the worst possible time. There will be rising military tensions between two nations, there will have just been an earthquake, the stock market will be in meltdown, or there will be a currency crisis. Food prices are escalating with climate change, and the global economy is already contracting under the weight of the huge debt pile we have accumulated over the last 40 years.

Further, it will definitely be argued that the timing is unfair to several sections of our global populace and that we must wait to allow another billion to securely attain middle-class status. This is rational, but it is wrong because it doesn't allow for the immediate risk that we face, the decimation of species happening this very minute, the opportunity that we have to improve our destiny or for the brilliance of our response.

It will always, always, be argued that it is the wrong time to form the GPA.

There may well be some modest success in one or two fields of global environmental protection, perhaps the first global carbon tax adhered to by a majority of the nations or a declaration that 15% of the high seas are inviolable which will be used to argue the status quo of using the fractured nation state system for biophysical protection. It is important to recognize these as being insufficient, that the biosphere will remain terribly exposed and the three friends of failure will still be in place. There must be regulatory and monetary power given to one specialist authority who can make rapid decisions, has the power of enforcement and is capable of ultra long term protection of the asset.

My salient test is a simple one: in my mind's eye I ask 'Have you got it? Have you got the oceans, the troposphere, the animal populations, the rainforests, the insects, the chemical testing, the landfill, the soil?' Until someone says 'Yes Angus, we've got them, we are looking after all of them, and that is our one and only job, 24 hours a day, decade after decade.'

Until then, our governance structure is not good enough. It must be global specialist protection of a global asset.

Because the formation of a global authority is a new concept and part of every one of us is a little wary of such large-scale change, it is natural to assume the worst and that everything will be difficult and opposed. But with the global citizenship voting the GPA into existence will come cooperation and positivity that will be extraordinarily strong. Of course, there will be obstacles to be overcome by the GPA, but they will be overcome.

A line in the sand has to be drawn at some point.

We must shoulder our obligation properly at some point.

That time is **now**.

Our first objective is that 150 million of us indicate that we are willing to vote.

You're lying in bed and having a restless night and you're mulling over the GPA in your mind. Or you're about to go online and register at **votegpa.com** or another such entity, or send a social media message advocating the formation of the GPA. Or perhaps you are about to talk to a friend or neighbour about how you would vote for a GPA.

And you hesitate.

Even though you know it's the right thing to do, that species are being destroyed this second, that the rainforest is disappearing, that plastic is pouring into the oceans, our soil is being scraped off, that the world is facing the real possibility of uncontrollable climate change, that there's no rational argument against us forming the GPA, that we have the numbers, the connectivity and that we can do it, a voice inside your head, deep down, says: "It's just impossible… isn't it?"

We only have our votes, our solidarity and a goal, and out there in the real world are armies, fighter planes, guns, nuclear warheads, a violent global drugs trade, brave journalists and environmentalists being murdered, existing institutions that are quite well resourced, and nation states run by predominately aggressive old men accusing each other of geographical boundary violations and embarking on yet another arms race.

I hear you. It can be an overwhelming thought, but we must remain focused on our above-all-others intergenerational responsibility, be brave and take the first steps, however small.

"One day you finally knew what you had to do, and began, though the voices around you kept shouting their bad advice. The generations before you failed. They didn't stay up all night. Nature beckons you to be on her side. You couldn't ask for a better boss."

Paul Hawken

My first talk advocating global governance of the biosphere was in a church hall, and I was nervous. You could throw a blanket over the audience, their number was so few. I think the attendance temporarily swelled by 20% when a stray dog wandered in. But I took a deep breath and just started and soon it was easy and the audience was agreeing.

The action we are about to take is just a change in human organizational form. And that means it is possible.

This is just us, just humans on this planet Earth – humans, as Yuval Noah Harari reminds us, who brilliantly collaborate with large numbers of strangers every single day. Just humans who, I'm sorry to say, will all be dead in 50, 60, 70 or 80 years' time.

But here we are now, at this point of time, this point of human history, standing on the shoulders of our predecessors who struggled, sacrificed, fought and worked to improve the human condition. Now it is our turn to reimagine and to make a key radical change for the greater good.

Most of the countries that we see today didn't exist just 200 years ago, so what's one supranational organization among friends? And we are friends, five billion of us connected and in the one global village. We are an unstoppable force for the change that must occur.

By setting a precedent, our first successful incursion into global governance may well have further positive ramifications. But for now, we must act with singular purpose: protect the biosphere.

The generations to come are watching us this minute. They are asking us, what **will** the biosphere look like in 500, 1,000 or 2,000 years' time? What did you leave us? What structure did you set up to ensure its health, when you knew the consequences of inaction and had the potential to act?

"You don't have to see the whole staircase, just take the first step."
Martin Luther King Jr

It is true that not every step of the whole staircase is visible to us, but let us take the vital first step together, to hold the vote and create a GPA.

History will judge that we just created another successful governance structure, another function of utility, and that we were absolutely right to do so.

We can form the GPA today with just one click on our phones, and I believe we will, because it is rational and urgent for us to do so.

1.5bn votes. One planet saved.

Thank you.

ACKNOWLEDGMENTS

My thanks to all those who have joined the journey so far, in particular Amber and Cindy, and to the special individuals who have formed the GPA working group. My sincere thanks to Chris, Darcey, Annabel, Paul, Charles and Tim for your advance endorsements of this book. Thank you to all at LID, especially Susan Furber who has steered me through a material editorial process. There is a better way to run a biosphere, let's go!